DDT

SCIENTISTS,
CITIZENS,
AND PUBLIC
POLICY

Thomas R. Dunlap

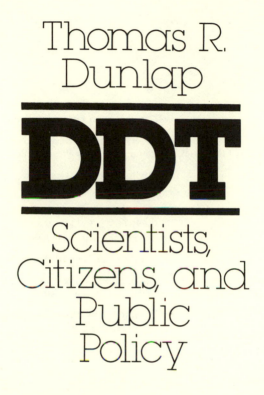

DDT

Scientists, Citizens, and Public Policy

PRINCETON UNIVERSITY PRESS

FOR MY PARENTS

ISAAC RICHARD DUNLAP
1908–1973

AND

FLORENCE RHINEHALT DUNLAP

With Love

CONTENTS

DDT

SCIENTISTS,
CITIZENS,
AND PUBLIC
POLICY

INTRODUCTION

EVEN THOSE who vividly remember World War II probably
retain only fragmentary memories of the introduction of
DDT, despite the wide publicity over the chemical. First used
on a large scale in the Naples typhus epidemic of 1943–1944
and during the rest of the war to protect millions of soldiers
and civilians against insect-borne diseases, it came home in
1945 on a wave of publicity and high hopes. It was the atomic
bomb of insecticides, the killer of killers, the harbinger of a
new age in insect control. Scientists predicted better and
cheaper control of agricultural pests, the eradication of imported insects—even, some thought, the end of insect-borne
diseases. A few towns used DDT during polio scares.[1] Still,
in 1945 technical marvels were part of daily life, and during
the postwar years DDT faded from public notice. Although
within a few years production of the chemical was far above
that of any of the earlier insecticides, few Americans paid any
attention. They saw fruits and vegetables with less insect
damage, but perhaps the only time they thought of DDT was
when they noticed the neighborhood spray truck going its
rounds in the summer.

By the late 1950s some scientists and citizens had become
concerned about the deaths of birds from concentrated sprays
used against the insect vector of Dutch elm disease, but most
Americans ignored DDT until 1962, when Rachel Carson's
Silent Spring appeared. The book touched off a heated debate,
for Carson, dissenting from the common view that DDT and
similar compounds were harmless, vigorously attacked both
the chemicals and the experts who recommended them. "It is
not my contention," she wrote, "that chemical insecticides
must never be used. I do contend that we have put poisonous
and biologically potent chemicals indiscriminately into the
hands of persons largely or wholly ignorant of their potentials
for harm"[2] The economic entomologists who recommended

the chemicals, the companies that made them, and other people—some scientists, some not—leaped to defend DDT and to condemn Carson as a crank. DDT became a political issue, as President Kennedy asked his scientific advisors for a report and Congress held hearings on federal regulation of pesticides. Although there was no immediate change in pesticide policy, *Silent Spring* permanently changed the climate in which the policy would be made. Pesticides were now a public issue and, through the 1960s at least, public concern over the environment would continue to increase.

While the public debate went on scientists continued to investigate DDT's effects in the environment and conservationists to protest its use. Late in the 1960s the focus of the anti-DDT campaign shifted from public education to litigation, as a new group, the Environmental Defense Fund (EDF), sought to bring the scientific evidence to court, before an "impartial arbiter," avoiding what it saw as an unresponsive bureaucracy and a lethargic public. The litigation, although extremely important, was also technical, and most citizens probably were as confused as enlightened by testimony about parts per million and billion, possible confusion of DDT and PCBs in the VPC, and the reproductive rates and migration counts of hawks they had scarcely heard of. If the scientists had agreed, it would have made a choice easier, but there were eminent authorities on both sides. Some doctors and medical scientists said that DDT was a menace to human health, others said that it was safe. Entomologists came to the witness stand to say that DDT was vitally needed; other entomologists appeared to say that it was not needed at all. Wildlife biologists claimed that DDT caused serious and possibly irreversible environmental damage; witnesses for DDT said that there were other explanations and that, in some cases, scientists had exaggerated the problems.

The environmental movement, in part spurred by public fears about DDT and other pesticides, provided a complement to the laborious, formal courtroom hearings. In the late 1960s groups demonstrated against pollution and for a clean environment and clean energy. Congress passed a variety of laws, including the National Environmental Policy Act (NEPA), and President Nixon found it necessary to remove his Secretary of the Interior, who had become too much of an environmentalist for the administration. In 1972, after a very long hearing, the Environmental Protection Agency (EPA) banned DDT—and faced appeals from the pesticide manufacturers, lumbermen, and interested citizens. Now, almost a decade after the Environmental Protection Agency's ruling, most people's memories about the controversy are very vague. Few remember the basis for the ban (except that it had something to do with wildlife and something to do with cancer), and many are still puzzled about the wisdom of the action. Some people remember the Environmental Defense Fund (EDF), which was the major force behind the campaign, but all too many confuse it with the Environmental Protection Agency, a government bureau (which pleases neither group). Why, after all, should anyone care about DDT?

This monograph, an attempt to place the controversy over DDT policy in historical perspective, grew out of the author's conviction that the episode was not merely startling but significant, that a history of the involvement of scientists and citizens in the making of policy on this issue could shed light on the social history of science in modern America. It is not simply that the consequences of DDT use are still with us in the form of residues in fat and breast milk, or that the kind of environmental contamination found in the one case has been confirmed for other chemicals. Both of these are valid reasons for an assessment of the controversy, for an effort to place it in the history of modern America. But the confrontation also brings to light the complex relationships among government agencies, the land-grant universities and their associated experiment stations, private industries, and the various sciences. It can tell us something about the nature of American support for science and the ways in which the social, economic, and political context in which scientists worked affected both the research they did and the ways in which their knowledge was or was not used in making policy.

The story of DDT policy begins well before the introduction of the chemical, for by the time it was discovered there already existed a system of regulation, based largely on experience with the arsenicals. Scientists had a set of criteria by which to judge both the efficiency and the possible dangers of insecticides and they were well aware of the political and economic limits on policy. Their expectations helped to guide research on the properties of the new chemicals and regulation of their residues on food. Chapter 1 deals with the development of economic entomology and the involvement of economic entomologists with insecticides. Chapter 2 recounts the work done by physicians, medical scientists, and government officials on pesticide residues as health hazards and traces the development of federal regulation in the pre-DDT period—the kind of scientific study done on the effects of residues, the assumptions about what was necessary to safeguard the public, and the nonscientific factors that limited the range of acceptable regulation.

The next section, Chapters 3 through 6, describes the early federal regulation of, and research on, DDT, the gradual discovery of the properties that many thought made it a threat to the environment, the way in which the public entered the debate, and the formation of a coalition of scientists, concerned citizens, and lawyers to mount a legal challenge to DDT and force changes in pesticide regulation. During the first decade of DDT use, research on DDT and controls on its use were largely governed by assumptions about the expected hazards of insecticides that had been derived from experience with the arsenicals. The only new group of scientists who became concerned with pesticides during this period were wildlife biologists, for DDT was useful in forests, swamps, and urban areas—places in which cost and toxicity had excluded the older pesticides. The studies of these scientists were focused on the immediate damage to susceptible species of heavy or misdirected sprays; only in the late 1950s did new evidence lead them to consider the long-term effects of low concentrations and to reexamine the behavior of DDT residues in the environment and in nontarget species.

Then, in 1962, Rachel Carson's *Silent Spring* converted what had been a quiet, scientific discussion into a noisy, public debate. Carson had done no original research, nor was her book a solid scientific synthesis of the data on pesticides in the environment (a better treatment, from that point of view, is Robert Rudd's *Pesticides and the Living Landscape*, published in 1964). Carson, though, did not intend to write a scientific work—she intended to warn and alarm the public and she succeeded brilliantly.[3] In the next few years, scientists in Europe and America pooled the scattered evidence against DDT and constructed the scientific case that increasingly alarmed citizens were to use in their legal battles. In 1967 a group of people on Long Island formed the Environmental Defense Fund, dedicated to the use of litigation in defense of citizens' right to a clean environment.

What may seem a disproportionate amount of space— Chapters 7 and 8—is taken up with the Wisconsin DDT hearing of 1968–1969, which was the first major legal challenge to DDT, and with the Environmental Protection Agency's hearing in 1971–1972. The record of the hearings is important for several reasons. First, it shows as no other source can, how the environmentalists gathered scattered evidence, pieced together the work of specialists in various fields, and integrated the whole into a logically satisying (at least to them) case. Second, the study of the opposition's arguments shows not only who publicly defended DDT, but also the assumptions, scientific base, and strength of the pro-DDT case. Finally, the hearings are one of the most significant developments of the DDT controversy—the use of litigation based on scientific testimony to give dissident groups an influence on policy. The techniques were not particularly new—the civil rights movement had an obvious influence on tactics—but the cases were an important step in the formation of the field of environmental law, and other groups and causes have adopted the methods used by the environmentalists to challenge other policies and the bureaucracies that made them.

The final part of the book describes the end of DDT and the beginning of the EDF's major drive against toxic chemicals in the environment—a campaign that grew out of the DDT episode. It discusses, as well, the general problem confronting Americans in the twentieth century—of which DDT is only the most famous example—the difficulty of using science to assess the costs and benefits of new technologies. Since the early twentieth century we have relied on scientific bureaucratic agencies vested with regulatory powers to safeguard the public from the hazards of new technologies. Despite lapses, or alleged lapses, of the system—asbestos, saccharin, cyclamates, thalidomide—the system continues, and critics of the agencies continue to concentrate on alleged improper relationships between the regulators and the regulated or on the agencies' failure to keep up with new developments. The hazards that came from widespread DDT use, though, were not entirely due to lack of scientific foresight, certainly not to any "shady" involvement between government officials and the chemical companies. They resulted from the policies of the Department of Agriculture (USDA) and of the other users of DDT, the result of an institutional and social framework that had grown up over the preceding half century. That the framework favored certain groups and certain ideas was both normal and natural, and it is difficult to see how it could have been otherwise; neither scientific research nor public policy are made in a vacuum or for a general "public interest," a conclusion that has many implications for Americans grappling with regulation and control of the byproducts of our industrial society.

My thanks to those who helped me in the preparation of this book must be widely distributed, for I began research on DDT in the fall of 1970, during my first semester as a regular graduate student in history at the University of Kansas, and finished revising this manuscript in the summer of 1979, after four years spent teaching at Virginia Polytechnic Institute and

State University. I became a graduate student in history after an undergraduate career and two and a half years of graduate study in chemistry—not ideal preparation for writing—and my graduate advisors deserve my special thanks. William M. Tuttle, Jr. of the University of Kansas set me on the track. He suggested that the early history of DDT might be an interesting topic for a master's thesis and then patiently set about teaching me to organize and write the paper, which turned out to be largely about the pre-DDT period. I went to the University of Wisconsin, Madison, in the fall of 1972, where Paul Glad continued the stream of advice and criticism while directing my Ph.D. dissertation. Paul, like Bill, was friend as well as mentor; I owe them both incalculable debts. My thanks, too, to my other teachers and the remaining members of my committees, especially Aaron Ihde, of the University of Wisconsin, Madison, Departments of Chemistry and History of Science, who virtually served as a second dissertation director.

My wife and colleague, Susan Miller, who never typed a page for me, has my special thanks. She has been a relentless and perceptive critic of my work, a patient auditor, and an invaluable support. James Whorton, author of *Before Silent Spring*, gave me an advance look at that manuscript and the benefit of his knowledge of pesticide regulation before DDT. John Perkins has furthered my education in the history of entomology and thoroughly criticized this manuscript; I could not have asked for a better reader. Douglas Helms of the National Archives and Records Service found a lot of material for me and discussed his own work on the boll weevil. Neil Larry Shumsky, one of my colleagues at Virginia Tech, read and criticized some of my work.

Many people supplied information. Mr. Walter Scott, then assistant to the secretary of the Wisconsin Department of Natural Resources (DNR), made available his private collection of materials relating to the Wisconsin hearing, which made my research much easier. Maurice Van Susteren, chief hearing examiner, loaned me a copy of the transcript of the Wisconsin hearing and spent several evenings discussing it with me. Charles F. Wurster, Jr. talked to me, wrote several letters and criticized an early draft of a paper I wrote on the Wisconsin hearing. Victor Yannacone gave similar aid, including access to his collection on the early work of the Environmental Defense Fund. The EDF office in Washington made available the legal papers and transcript of the Environmental Protection Agency's hearing on DDT. Others who wrote letters or granted interviews were: Donald A. Chant, R. Keith Chapman, Roland Clement, Arthur Cooley, Francis B. Coon, Ellsworth H. Fisher, Myra Gelband, Theodore Goodfriend, Whitney Gould, Wayland J. Hayes, Jr., Joseph J. Hickey, Hugh H. Iltis, Lorrie Otto, Louis A. McLean, Marvin Merta, Dennis Puleston, William Sax, Willard Stafford, Anthony S. Taormina, George J. Wallace, George Woodwell, and Carol Yannacone. The Wisconsin Legislative Reference Bureau sent me the drafting file on the DDT bills and information on the DDT ban in Wisconsin. My thanks to Sage Publications (London and Beverly Hills) and to *Wisconsin Magazine of History* for permission to reprint material that appeared in slightly different form: "Science as a Guide in Regulating Technology: The Case of DDT in the United States," *Social Studies of Science* 8, no. 3 (August 1978), 265–285, which is incorporated into Chapter 4, and "DDT on Trial: The Wisconsin DDT Hearing, 1968–1969," *Wisconsin Magazine of History* 62 (Autumn 1978), 3–24, now parts of Chapters 7 and 8. I am very grateful to my editors at Princeton University Press—Gail Filion, Margaret Riccardi, and Cathy Thatcher—for their aid. Finally, a National Wildlife Federation Fellowship supported me during my last year in graduate school and Virginia Tech provided small research grants.

ABBREVIATIONS

AAEE—American Association of Economic Entomologists, the most important professional organization of economic entomologists. Charles V. Riley, then head of the U. S. Department of Agriculture, Division of Entomology suggested the idea of the organization, and his assistant, Leland O. Howard, and James Fletcher, Dominion entomologist (Canada) wrote the constitution. Riley presided over the first organizational meeting, 1889, and was first president. Although the group quickly dropped the requirement that members hold official posts as entomologists, government entomologists continued to dominate it and, until the founding of the *Journal of Economic Entomology* in 1908, the USDA published the records of the organization.

BEPQ—Bureau of Entomology and Plant Quarantine. The original Federal office for entomology was a bureau of agriculture in the patent office; in 1853 it hired an entomologist, Townend Glover. Riley succeeded to the office in 1878—it was then part of the U. S. Department of Agriculture—and, except for two years (1879–1881) held the position to 1894. His assistant and successor, Leland O. Howard, was chief of the bureau until his retirement in 1927. The Division of Entomology became a bureau in 1904, and in 1934 the Bureau of Plant Quarantine and various other offices were added and the BEPQ came into existence. In a reorganization (1953), the bureau was split up and lost its distinctive name. Many of its functions were placed in the entomology research branch of the U. S. Department of Agriculture's Agricultural Research Service (ARS), and (since 1978) the Science and Education Administration.

CNRA—Citizens' Natural Resources Association of Wisconsin, Inc. A conservation group, the CNRA was active in

11

the fight against DDT in Wisconsin during the 1960s and brought the petition for the banning of DDT that led to the Wisconsin DDT hearing of 1968–1969 before the Department of Natural Resources.

DDT—D(ichloro)d(iphenyl)t(richloroethane). See Appendix B.

DNR—(Wisconsin) Department of Natural Resources, the state agency responsible for the conservation of Wisconsin natural resources and, insofar as the state articulated a policy, for the preservation of the natural environment. The DNR became involved in the DDT action because of its legal responsibilities to enforce the antipollution laws of the state.

EDF—Environmental Defense Fund. Scientists interested in the preservation of the environment, other citizens, and a lawyer formed this Long Island based group in October 1967, following the successful action against the Suffolk County Mosquito Control Commission, *Yannacone v. Dennison.* Several members of the group had already met and worked together in the Brookhaven Town Natural Resources Council, another citizen group.

EPA—Environmental Protection Agency. Formed by Reorganization Plan Number 3, this agency was created to enforce environmental laws and regulations. It received statutory authority over pesticide regulations, which is shared with the Department of Health, Education, and Welfare, in December 1970, when Congress largely stripped the U. S. Department of Agriculture of regulatory power in this area.

FIFRA—Federal Insecticide, Fungicide, and Rodenticide Act, passed, 1947. It provides the legal basis for registration and regulation of pesticide chemicals. It was supplemented by the Federal Environmental Pesticide Act of 1972, which extended federal authority to intrastate uses of registered chemicals.

FDA—Food and Drug Administration. See Appendix D—Enforcement of the Food and Drug Laws.

GAO—General Accounting Office.

HEW—Department of Health, Education, and Welfare; now Department of Health and Human Services.

IPM—Integrated Pest Management, the practice of pest control through the use of various methods, including chemicals, to reduce pest populations below the level at which they cause economic damage to the crop.

NACA—National Agricultural Chemicals Association, a trade association of corporations manufacturing agricultural chemicals. The group first assumed an important role in lobbying for pesticide tolerance levels and regulations after World War II. The trade journal *Agricultural Chemicals* was the voice of the industry. The association's Industry Task Force for DDT took responsibility for coordinating the industry's defense of DDT in the wake of *Silent Spring*.

NEPA—National Environmental Policy Act. Passed by Congress in 1969, signed into law 1 January 1970, it sets out a national policy to preserve the environment for future generations and (Section 102) requires an environmental impact statement for major federal projects that affect the environment.

PCBs—P(oly)c(hlorinated)b(iphenyls). There are a series of compounds of similar structure, all polychlorinated biphenyls, which are known as PCBs to the industry. Manufactured for use in electrical equipment, the compounds were spread widely throughout the world and have leaked into the environment. Although themselves condemned as environmental pollutants, their importance in this narrative is that they were confused, or allegedly confused, with DDT and its metabolites in gas or vapor phase, chromatography. See Appendix G—Vapor Phase Chromatography.

USDA—United States Department of Agriculture. It was formed in 1862 and assumed cabinet-level status in 1889. Federal entomological work is contained in this agency and, through its connections with the land-grant schools, under the Hatch Act of 1887 and the Adams Act of 1906, it exercises a degree of control and direction over state work. Until 1940 it also contained the agency responsible for the enforcement of the food and drug laws, a situation that led it into a conflict of interest.

VPC—Vapor Phase Chromatography; see Appendix G.

WARF—Wisconsin Alumni Research Foundation, the organization that conducted much of the analytical work for Hickey and others on pesticide residues in the environment.

Part One

PREPARATION FOR DDT

CHAPTER 1

Economic Entomology
and Insecticides

Stand dismayed, you farmers
wail, you vinedressers,
for the wheat, for the barley;
the harvest of the field has been ruined.
Joel 1:11

Spray, farmers, spray with care,
Spray the apple, peach and pear;
Spray for scab, and spray for blight
Spray, O spray, and do it right.
E.G. Packard[1]

IN ORDER to understand the controversy over DDT, one must first see it in context. In 1945 the "atomic bomb of insecticides" was a novelty, but chemical insecticides were well established, to the point that they were almost synonomous with insect control.[2] Farmers, economic entomologists, and government agencies had had a half-century of experience in using, recommending, and regulating insecticides. They had a frame of reference into which they would easily fit the new chemical and others like it. It was this frame of reference, a set of unspoken and almost unconscious assumptions about the need for, and the uses, disadvantages, and possible dangers of chemical insecticides, that accounts for the enthusiastic and almost uncritical acceptance DDT found in 1945, and that helps explain both the wide use and the passionate defense conducted later.

The frame was assembled in the late nineteenth and early twentieth century, as economic entomology and chemical insecticides grew out of efforts to meet the problems of insect infestation that plagued the American farmers in this period.

Although these were years of enormous agricultural expansion, they were troubled times. Falling crop prices, high railroad rates, and natural disasters made farming an often frustrating business. Among the difficulties was an increase in insect damage to crops, a problem one historian has labeled the "insect emergency."[3] So it must have seemed to the inhabitants of the Great Plains and, to a lesser extent, to the rest of the country. In the 1870s a series of grasshopper swarms ate up the crops. They were followed by the chinch bugs, which ruined farming in the eastern edge of the Plains. The army worm, the cotton worm, the codling moth (apple worm), Hessian fly, Colorado potato beetle, and a host of other pests devoured the crops. Even in Golden California various scale insects threatened to wipe out the young citrus fruit industry.[4]

The insect emergency was in large part due to the progress of the period: the spread of commercial agriculture across the continent, the regional specialization that was coming to mark production for a national market, and the increased speed and volume of transportation meant that diverse ecosystems were being replaced by simpler ones, by continuous areas of food and shelter for the insects that ate the crops, and that railroads and wagons were busily (if unintentionally) spreading pests to these new and attractive homes. An excellent case in point is the Colorado potato beetle. In the early 1860s wagons returning from the Colorado Rockies brought the beetle from its native habitat, where it lived on a variety of plants, to Illinois. There it found ideal conditions, enormous supplies of food (it quickly came to favor the potato vine), few natural enemies adapted to prey on it, and relatively continuous cultivation. It spread from patch to patch, flying or traveling on wagons.

Native insects that developed a fondness for crops were only part of the farmers' problems. There were other insects that normally preyed on the crops and, in addition, many European pests accidentally introduced into the United States by steamship and clipper. These posed a particularly serious danger. Most American crops were native to Europe or Asia, where their pests had developed along with enemies to prey on them. Settlers had brought the crops and commerce seemed, inadvertently, to be bringing the pests, but there was no one to fetch the parasites and predators. As a result, insects that were not important problems in their homelands caused enormous damage here. The fluted, or cottony-cushion, scale, which threatened the California orchards in the late nineteenth century, was native to Australia. The gypsy moth came from Europe, the boll weevil from Central America (via Mexico). In 1897 the head of the USDA's Division of Entomology, Leland O. Howard, found that thirty-six of the seventy-two most dangerous insect pests in the country were of foreign origin and another six were suspected of being aliens. Worse, Howard warned, there were others, such as the Mediterranean fruit fly, which might enter at any time. He ended with a plea for an adequate quarantine inspection service.[5]

The transportation of insects around the world and their adaptation to new areas was a significant part of the breakdown of isolated floras and faunas that took place in the nineteenth century. American farmers, though, if one may judge from the literature, cared little about the biological significance of their problems. They just wanted to get rid of the pests, and in their desperation they turned to all kinds of remedies, from patent bug-killing machines and days of prayer and fasting (declared by the governor of Missouri during the "hopper" epidemics of the 1870s) to a bewildering variety of washes, emulsions, baits, and sprays. The favorite remedy, which gained in popularity during this period, was chemical poison.[6] The first chemical used on a large scale was paris green—a common pigment for paints and wallpaper—and its first target was the Colorado potato beetle. Beyond these few facts, it is difficult to trace the early history of insecticide use. There were several claimants to the honor of inventor of the beetle poison. In 1868 a farmer in Fond du Lac, Wisconsin, said that an Iowa man had told him that paris green mixed with ashes and sprinkled on the vines would poison beetles. The same year J. P. Wilson patented a solution of one part

paris green to two parts mineral oil for the same purpose.[7]
Other stories circulated through the rural press, along with
recipes. It is probable that several people independently made
the discovery. Paris green was, after all, copper aceto-
arsenite, and everyone knew, if he knew nothing else about it,
that arsenic was a deadly poison. What more natural than to
put it on the vines and see what happened?

The use of chemicals increased, slowly at first, then faster
and faster. Farmers tried paris green on other crops, tried
other arsenicals, and worked on sprays and spraying equipment. Mixtures of lime, sulfur, and salt became popular for
orchard sprays and kerosene emulsions found favor against
sucking insects. By the late 1880s the newly established
agricultural experiment stations, established under the Hatch
Act of 1887, were giving the farmer professional advice, and
by the end of the century a small but growing insecticide
industry was both supplying materials and sending salesmen
into the field to tell the farmer how to use the product.
Chemical control was becoming an accepted weapon against
almost all insect pests.[8]

The increasing use of insecticides coincided with the rapid
development of economic entomology under the auspices of
the government. Although the federal government had employed an entomologist, Townend Glover, since 1853, and
various states had employed entomologists, there were no
groups that could call themselves economic entomologists;
nor did the lonely practitioners have much public recognition.
Indeed, when swarms of grasshoppers swept over the Great
Plains in the 1870s and the westerners demanded relief,
Congress completely ignored Glover and the Department of
Agriculture. Instead, it appropriated $18,000 for an investiga-
tion by the Geological Survey. The Survey promptly chose
two of its former agents, Cyrus Thomas and A. S. Packard,
who had done entomological work, and filled out the new U.
S. Entomological Commission with the state entomologist of
Missouri, Charles V. Riley.[9]

The commissioners knew of no remedy for the grasshopper
outbreaks, and their only service to the community was to

predict the extent and duration of the infestation, a service
Riley had done for Missouri. Its most important contribution
was to establish Riley in the federal service. Glover retired in
1878, the Entomological Commission was transferred to the
USDA, and Riley became head of the Division of Entomology. Except for two years (1879–1881) he headed the division
until 1894. Though he was, like Glover, a self-trained naturalist and entomologist, he was a very different sort of
administrator. Where Glover had been content to occupy a
niche, Riley set out to build an empire. His assistant and
successor, Leland O. Howard, described him as a "restless,
ambitious man, a great schemer, and striving constantly to
build up a large organiztion."[10] Riley's schemes, in fact,
helped ease him out of government. His constant lobbying,
attempts to build the division, and political connections (by
which he sought to preserve his position) eventually led to
clashes with the Secretary of Agriculture and to his resignation in 1894. Before he left, though, he laid the foundations
on which his former assistant built.

Howard became the chief, a position he retained until his
retirement in 1927. He was the real empire builder. Under his
leadership the bureau (it attained that status in 1904) became a
respected and established part of the government, and he
played a large part in making economic entomology a professional applied science—a group with standards, training,
entry requirements, and a defined public mission recognized
by its practitioners and the public alike. The process was
intimately connected with the rise of chemical insecticides and
with the establishment of the institutions and bureaus that
played a crucial role in the rapid and widespread use of DDT
after World War II.

When Howard took office there was little formal training
for economic entomologists. They were usually naturalists
who had taken a scientific interest in the problem. Riley, for
example, had been raised and educated in England and had
come to agricultural entomology only when he arrived in the
United States and, at the age of seventeen, began work on a
farm west of Chicago. Others of his generation came from

disparate backgrounds. Cyrus Thomas had been first a lawyer, then a minister. In 1869 he had become an entomologist and botanist in the Geological Survey of the Territories and, for a short time, a professor of natural history. He eventually abandoned entomology for ethnology and took a position with the U. S. Bureau of Ethnology. John Henry Comstock of Cornell, who had been head of the Division of Entomology for two years (1879–1881), C. H. Fernald, credited with first using lead arsenate as an insecticide, Stephen A. Forbes, president of the American Association of Economic Entomologists (AAEE) in 1893 and 1909, and A. J. Cook of Michigan, a pioneer in both arsenicals and kerosene emulsions, were "distinctly self-trained and self-educated in entomology."[11] Howard was professionally trained, but he had been Comstock's first student. By the 1890s, though, the pioneers of the profession had the raw material at hand to form a professional group. The Morrill Act of 1862 had created a system of higher education that the Department of Agriculture could shape to its purposes, and the Hatch Act of 1887 provided agricultural research stations at each college. The personnel of the station were both employees of the Department of Agriculture and faculty members of the University, an interesting if somewhat vexing situation,[12] and the federal government, through an annual appropriation, supported the station and had some control over the research. The Act changed the department from a small agency in Washington to a "nexus of semi-autonomous research institutions permanently established in every state" and provided the first job market for economic entomologists.[13]

Before he left, Riley had taken the first tentative steps to weld this group together. In the January 1889 issue of *Insect Life,* a Department of Agriculture periodical, he suggested forming a professional organization of government entomologists. That summer Howard and James Fletcher, the Dominion entomologist (Canada), worked out a constitution, and there was an organization meeting in August at the convention of the American Association for the Advancement of Science in Toronto. Riley was chosen president.[14] The new group, the Association of Official Economic Entomologists, at first limited membership to those who held positions with the government or an agricultural experiment station. It soon dropped the requirement and the word "official" from its title, but it retained ties to the government and drew most of its members from the ranks of official entomologists. Until the society began the *Journal of Economic Entomology* in 1908, the Division of Entomology published its proceedings, first in *Insect Life*, then in the bulletins of the division. Most of the active members, and almost all who presented papers, were official entomologists, and these men dominated the officers of the association.

When Riley stepped down, though, the division was still groping for a mission. Rather than looking for problems to solve, it let them develop and waited for others to seek its advice. It had a minute budget (about $20,000 a year), a handful of people, and little prestige. The process of change and development that made the Bureau of Entomology a secure and established agency in the next twenty years followed a path others also trod. In *Science and the Federal Government*, A. Hunter Dupree outlined the general characteristics of the scientific bureau that grew up in the period after the Civil War, features apparent in the new Bureau of Entomology.[15] A scientific agency, Dupree said, was problem centered, not discipline centered. It did not seek to advance a particular scientific discipline on a broad front, it tried to apply the techniques and knowledge of the discipline to a particular set of important, public problems. Each bureau sought to establish itself as the recognized expert on certain problems and to realize some continuity, ideally through a Congressional grant of power, for continuing regulatory or administrative functions.

The ideal bureau chief sought other goals as well: control over the budget, a loyal corps of workers, and a stable set of outside alliances. Some agencies, such as the Public Health Service and the Forest Service, put their employees into uniform; most, though, built morale and loyalty by fostering a sense of common purpose. The ideal bureau chief cultivated strategically placed Congressmen, sought connections with universities, which could provide new recruits for the agency, and with professional groups. And, in all cases, the bureau sought to serve some personal and interested segment of the population that would provide political support.

The Bureau of Entomology fit this pattern quite well. It had a field in which it could, and did, establish itself—the study and control of insects injurious to man, animals, and crops. The division of the bureau into groups organized around particularly important insects of types of crops reflects the emphasis on the application of knowledge rather than the development of entomology. Passage of the Insecticide Act of 1910 and the Quarantine Act of 1912 gave the bureau continuing responsibilities in agriculture as well as more power to carry out its mission of aiding the farmer. Its close association with the American Association of Economic Entomologists is evidence of its concern for outside allies, and the connections between the Department of Agriculture and the land–grant colleges and experiment stations provided a source of personnel, local agents to handle particular problems, and a further source of political aid.[16]

"By 1916," Dupree noted, "the metamorphosis of the Bureau of Entomology into a new scientific agency was virtually complete, and it was proving its worth so regularly that its position in government was not only secure but taken for granted."[17] The bureau had "proved its worth" by suggesting methods of insect control that would solve the farmers' problems, and the way in which it met its responsibilities led directly to the atmosphere in which DDT was uncritically accepted and widely (some would say indiscriminately) used. The bureau and economic entomologists in the experiment stations became committed to chemical control at the expense of other methods and, according to some entomologists, at the expense of the scientific discipline itself.

How did this situation come about? Why did the economic entomologist, whom Stephen A. Forbes described in 1915 as an "ecologist pure and simple, whether he calls himself so or

not. . . ." turn so far away from this study that another
founder of the discipline could, in 1924, complain that the
younger generation of scientists was "losing sight of the
insect"?[18] There is no simple answer, for individual entomologists tried and recommended other methods, and the
Bureau of Entomology continued work on cultural and
biological controls for years. The advantages that chemical
insecticides offered for the official economic entomologist,
though, played a major role. The economic entomologist
"justified his work and made his claim for federal and state
funds upon his ability to make sound recommendations for
the control of various insect pests,"[19] and insecticides, more
than their major competitors, provided control that was
satisfactory to the farmer, whose demands the bureau had to
satisfy.

Consider, for example, the bureau's experience with the
boll weevil, probably the most important pest to enter
the United States in the last half of the nineteenth century.
The tale of the weevil is an illustration of the reasons for the
general failure of cultural controls. Soon after the insect
crossed the Rio Grande, entomologists working for the
bureau devised a set of farming methods—a cultural control
system—that would check it and from 1895 to 1918 the
bureau conducted a vigorous campaign to persuade southern
cotton farmers to use these methods. Although it tried
experimental farms, demonstrations, and propaganda, and
although the weevil wreaked havoc throughout the cotton
belt, the bureau had little success. On the other hand, a
system of chemical control, offered in 1918, found immediate
acceptance, as farmers quickly adopted calcium arsenate and
the bureau's recommendations for its use. The answer to this
puzzle lies, not in science, but in economic and social arrangements.

The weevil story began with a letter from a farmer to the
Department of Agriculture. In October 1894, C. H. DeRyee
of Corpus Christi, Texas, informed the department that "a
peculiar weevil or bug which by some means destroys the
squares and small bolls" [of the cotton plant] had almost

destroyed the late cotton harvest.[20] Howard quickly hired C.
H. Tyler Townsend, an entomologist working in New
Mexico, to survey the infested areas in Mexico and Texas and
to suggest remedies. Townsend outlined a program to check
the weevil's reproduction during the summer, to limit its
food supply during the early fall (when it migrated), and to
reduce the over-wintering population. Clean cultivation and
widely spaced rows would, he said, reduce the weevil's
reproduction in the summer by creating unfavorable conditions for the larvae, which grew in the fallen cotton squares.
The key to the program was the destruction of the plants in
the early fall after the main crop had been gathered but before
the first frost, when the insect went into hibernation. Burning
or plowing under the stalks would deprive the weevil of food
just before the fall migration, when it needed a large supply,
and drive it into hibernation without enough to carry it
through the winter. Where practical, farmers should flood
their fields to kill the hibernating weevils in the ground. Such
a program would, Townsend admitted, eliminate any chance
of a "top crop," a late picking that would mature if frost were
delayed, but it would insure a much better crop the following
year.[21]

The bureau recommended cultural controls to Texas
farmers until 1898, when Texas appointed a state entomologist. It resumed work on the problem in 1900, when
the weevil crossed into Louisiana, and Congress made an
emergency appropriation. It attacked the problem from several sides, working on biological control, chemical insecticides, and the breeding of resistant varieties of cotton, but
there seemed no viable alternative to cultural controls. It
recommended a modified version of Townsend's program to
farmers and urged crop rotation, the abandonment of marginal lands, diversification of farming in the region, and
quarantine measures to slow down the spread of the insect
through the cotton belt.[22]

The program, though, required either community cooperation or legal force to be fully effective. Although an individual
farmer could derive some benefit from practicing the bureau's

program, it was better if an entire district cooperated. Unfortunately, farmers would not all practice cultural controls and strongly resisted the passage of laws that would have forced them to do so. Townsend had been sure that this would be the case when he had recommended a ban on cotton culture along the international border to stop the boll weevil. A no-cotton zone, he said, would be ineffective unless enforced by law. "Cotton is the only cash crop here," he wrote to Howard, "and the people will raise it unless compelled by law not to." He was equally certain that the Texas legislators would never vote for such a law because they feared their "constituents . . . [who] cannot see the matter in its true light." Even if the legislature passed a law, he gloomily concluded, it would be ineffective, for popular resistance would prevent enforcement.[23]

He was right; public pressure defeated every suggestion of a ban on cotton culture, even when the legislation was for a limited time and covered a limited area. In 1895 Governor Culberston of Texas, at the urging of Howard and the Secretary of Agriculture, proposed a fifty-mile wide no-cotton zone along the border with Mexico. The legislature promptly rejected the suggestion, and later attempts to combat the weevil by laws met with the same fate.[24] In 1904 Texas offered a $25,000 prize for an effective method of weevil control, and the governor appointed a board of farmers to judge the proposals submitted for the award. The board concluded that cultural controls were the only practical remedy. It recommended state laws enforcing the early chopping and burning of cotton plants. The legislature took no action. The same year a convention in Shreveport, Louisiana, called to consider the boll weevil problem, rejected a plan for a one-year moratorium on cotton growing in infested areas.[25] Even such mild measures as quarantine laws encountered strong opposition. Not until 1903 did any of the still uninfested southern states adopt a quarantine law, and not until 1906 did all of them have this protection.[26]

Farmers resisted cultural controls for a variety of reasons. For many, especially the tenant, the bureau's program meant certain losses but uncertain gains. Its program, the bureau admitted, "in some cases implies operations that would not be the proper ones in all cases for production of the largest crop were the pest not present."[27] Destroying the cotton plants after the main harvest, the most important part of the operations, meant that the farmer sacrificed any chance for a "top crop" that year. He would, the bureau said, be rewarded with a much larger main crop the following year, but many farmers were reluctant to take the chance. Weather, other pests, or natural disasters might ruin the entire crop next year, or the price might fall. Tenants, who were normally bound by yearly contracts and had no security of tenure, had even less incentive to practice the controls. Unless they were sure of being on the same plot the next year, they might be sacrificing their own interests for someone else's.[28] The program also had implications for the entire region. Crop rotation, diversification of agriculture, and abandoning marginal lands—all parts of a full-scale attack on the weevil— would have serious effects on the regional economy. The crop-lien system, sharecropping, tenancy, and an entire complex of economic arrangements centered on the annual cotton crop. Few were willing to undertake a major change in the agricultural economy unless forced to do so.

There were other, less tangible, reasons. Custom was a powerful force. Tenant farmers in Texas, for example, resisted chopping and burning the stalks in the fall, even after frost had killed all chances for late picking, because that job was traditionally part of the tenant's responsibility in the spring.[29] Farmers also had their own ideas about the best varieties of cotton, the best practices, and the best time to plow the stalks under as well as a strong suspicion of "book-farming." Even when farmers could be persuaded that the control measures would work, they were often unwilling to clear their fields. Their neighbors might not, and that meant, they thought, that their own efforts would be wasted.[30] Combined with the unwillingness to accept legal enforcement, this severely retarded the acceptance of cultural control.

Chemicals were the favorite method, and despite bureau experiments and the farmer's own experience showing that they were ineffective, farmers continued to buy and use them. As late as 1904 an entomologist estimated that Texas farmers used twenty-five carloads of paris green in futile attempts to kill the pest.[31] The feeling was deep rooted. Discussing reaction to the bureau's cultural controls in 1898 Townsend observed that "[m]any [farmers] do not take to the recommendations. . . . They want a poison to kill them quick and be over with it."[32] Even Texas officials were impatient with remedies that were "troublesome or indirect."[33]

Looking for a "quick fix," farmers tried a variety of methods, some quite bizarre. Light traps enjoyed a vogue, as did steeping cotton seeds in sulfur (to make the plant unpalatable to the weevil). Farmers invented and patented a variety of machines to kill the weevil; one sucked them off the plants and incinerated them. Other suggestions included sterilizing weevils with X-rays or electrocuting them by passing a current through the fields.[34] In 1904 O. F. Cook, a plant pathologist with the Bureau of Plant Industry of the USDA announced that he had found a small ant in Guatemala that fed on the weevil. Newspapers and the public hailed the discovery and the Bureau of Entomology with many misgivings spent more than $25,000 importing the ant, fighting an injunction to stop its use (brought on the grounds that it would be a pest itself), and raising colonies in Texas. The ant did not live up to its press notices. It was not especially adapted to prey on the weevil—in fact it hardly bothered the pest—and it failed to survive the Texas winter.[35]

Both before and after this fiasco the bureau had been interested in biological control. Townsend had recommended the method in 1894. The legislature would not pass laws enforcing cultural controls and farmers would not voluntarily undertake them. Parasites, which would work regardless of legislature or farmer, were essential.[36] Despite a good deal of investigation the bureau never found that such measures were particularly useful against the pest. Though there were many parasites, some of them American insects that also preyed on closely related weevils, none of them, or any combination, seemed to hold the population of the boll weevil below the level at which it would damge the crop.[37]

In an effort to find some acceptable means of control, the bureau continued to work on pesticides. The major problem had been that the insect spent its life either in the developing boll or the shelter of the bracts (shucks of the boll). Further work on the life history of the pest, though, showed that the adult sipped dew, and entomologists worked out a program to take advantage of this habit by dusting with finely powdered calcium arsenate early in the morning—which poisoned the dew. An initial treatment killed most of the over-wintering weevils as they emerged and later dustings eliminated enough of the later broods to keep the population below the economic threshold, the level at which the damage began to affect the crop yields.[38]

The bureau quickly began to tout calcium arsenate as the answer to cotton farmers' problems, asking rhetorically: "can the boll weevil be controlled profitably? . . . An affirmative answer to the question, eagerly sought ever since the weevil invaded this country, has at last been found."[39] In Louisiana, the extension division of the state university quickly issued four circulars describing the new system, the materials and equipment needed, and the method of checking for weevil infestation. Two of them concluded with the exhortation that "It does not COST TO DUST, it does COST NOT TO DUST."[40] Despite warnings that cultural controls were "absolutely necessary in any system of weevil control," farmers, the bureau, and the extension entomologists began to rely entirely on calcium arsenate,[41] and there was an immediate boom in its production. In 1918 the sole manufacturer had found it difficult to sell 50,000 pounds of his product; two years later twenty manufacturers were producing 10,000,000 pounds a year, and by 1927 a company formed to dust cotton from the air had contracts to treat 500,000 acres.[42] Through the 1930s production continued to climb, until calcium arsenate was, second to lead arsenate, the most popular of the arsenicals.

The boll weevil project shows clearly the nonscientific problems that economic entomologists faced; their chief obstacle was not to find a method of control, but to find a method that did not conflict with the social and economic needs and desires of the community. Changes that seemed both reasonable and simple, affecting only the farmers' work habits and to some extent his crops, turned out to have far-reaching consequences in the community. It was a difficulty that the bureau encountered in other areas as well. During the late 1920s, for instance, it attempted to combat the European corn borer by the same methods used against the weevil: destruction of its food and hibernation niches by destroying the stalks after the harvest. Despite a large campaign and state cooperation, including laws enforcing the measures, it encountered resistance. Even in an area where alternative crops were available, farmers resisted change, preferring to spend money on insecticides and to continue farming in the old way.[43]

Cultural control, though, was not the only alternative to chemicals. There was also biological control, the use of predators, parasites, or disease to check injurious insects. Between 1889, when they first demonstrated the usefulness of natural enemies on a large scale, and the 1930s, entomologists undertook several large projects of biological control. The campaign was stimulated by Riley's importation of a vedalia beetle from Australia to control the fluted, or cottony-cushion, scale. The scale had entered California in the early 1870s and quickly spread through the groves. Its ravages were so severe that by the mid-1880s Californians were in despair. Orchards were being ruined and there seemed no way to stop the pest. Neither arsenicals nor emulsions—of kerosene, oil, or soap—had much effect. Riley found that the pest was not a serious problem in Australia. He reasoned that it was native there and that natural enemies kept it in check. With the aid of a friendly California Congressman he sent one of his assistants, Albert Koebele, to Australia as a member of the American delegation to the Melbourne Exposition of 1889 (circumventing Congressional restrictions on the use of some

Department of Agriculture funds for foreign travel). Koebele sent back some specimens and returned with another shipment, including 128 specimens of a vedalia beetle that preyed on the scale. The beetle, which preyed on eggs, larvae, and adults, quickly brought the scale under control; by the end of the first year areas in which the beetles had been released were almost free of the pest. Enthusiastic farmers quickly spread the news, and the beetle, throughout the state.[44]

The success of the vedalia raised false hopes. Some extreme advocates of the method claimed that it was a panacea. All one had to do was to find the enemy of a pest (it was assumed that there was one), bring it to this country, and the job was done.[45] On the crest of this wave of popular enthusiasm, Congress and the state governments underwrote a series of investigations into biological control; but by 1920 the boom was over. The bureau's work against the gypsy moth, one of the major eastern forest pests, shows why.

The gypsy moth, like many other important pests, is not a native. A French astronomer, Leopold Trovelot, imported the moth in the late 1860s to use in his hobby—crossbreeding silk-producing caterpillars. He lost some of the eggs and the moth established itself in a vacant lot near his home in Medford, Massachusetts. It remained a local curiosity until the summer of 1889, when a population explosion brought a plague of caterpillars. In response to the town's plight, Massachusetts launched an almost successful eradication campaign. In 1900, though, the legislature decided that the moth was now a minor nuisance and that further appropriations were not needed. The moth quickly recovered—the summer of 1905 in Medford was much like that of 1889—and the bureau stepped in to aid the state.[46]

By 1905 the moth had spread too far to make eradication practical, and the entomologists turned to biological control. The state appropriated $30,000 for the work for the period 1905–1908 and funded some of Howard's travels to Europe to gather parasites. Congress appropriated $25,000 for the same purpose. The bureau set up a laboratory to care for the parasites until they would be released, and threw its resources into the project. At the outset both the public and the entomologists in charge of the work were sure that biological control would quickly bring the moth population down. Howard predicted success in two or three years, and public confidence ran so high that he had to warn against abandoning conventional methods of control. Although natural enemies would soon end the moth problem, any increase in the insect's numbers, he said, would only make the enemies' job harder and delay complete success.[47]

The program was not based on a careful study of the moth's ecological situation or even on an understanding of the place of natural enemies in the moth's population biology. Indeed, the ecology and population biology of the moth were almost completely unknown. Entomologists worked from their general knowledge of entomophagy and the moth's status as a minor pest in Europe, and counted on a repetition of Riley's success with the vedalia. The lack of detailed knowledge was not unusual; ecology was a very new science, indeed, it was still a part of biology and natural history and not a separate discipline at all. Economic entomologists, though, could not afford the luxury of waiting for full information about the moth and its enemies or for a fully developed theory of ecology before proceeding to control. Problems (and the public) demanded solutions, and the scientists had to do the best with what they had.

Unfortunately, they were to find that the California experience had been quite exceptional. Establishing parasites and predators in a new environment proved to be much more difficult and expensive than anyone had anticipated. Some of the parasites died, others refused to breed, still others vanished without trace when released. Even those enemies found preying on the moth in the field seemed to do no good. By 1908 both Howard and W. F. Fiske, head of the Melrose Highlands laboratory, were discouraged about the prospects of using the moth's enemies to control it. In April Fiske wrote gloomily that he now thought that the work would begin to show results about the time of Philippine independence. Howard replied that he had previously thought that three years would have been enough, but now he thought more time would be required. In another letter, to a member of the Boston Chamber of Commerce, he suggested an outside limit of seven years before any results would be apparent. The problem, he said, was "extremely complicated;" in Europe fifty parasites combined to control the moth and success here might depend on transplanting all of them to the United States.[48] In a long technical report, prepared in 1910, Howard and Fiske told why they now believed that, for the foreseeable future, natural enemies would not be able to control the moth's population surges. Successful control, they said, depended on a thorough understanding of the parasites' environmental requirements, and this information simply was not available. The Bureau, they said, should continue the project, but not with the expectation that the problem would soon be solved or that it would be done cheaply. Conventional methods would still be needed.[49]

The project had been a stunning scientific success; entomologists had vastly increased their understanding of the moth, of its enemies, and of the possibilities and limitations of biological control. The public, though, was less interested in scientific advances than in relief from defoliation and the nuisance of caterpillars. Massachusetts, which had supported biological control, quickly stopped all funds for the parasite laboratory and concentrated on chemical and mechanical controls in towns and along roads. Although the bureau continued to import parasites until World War I severed its connections with Central Europe, it did not emphasize the project. It concentrated on charting the limits of the infested area and advising state authorities on control measures.[50]

The problem with the gypsy moth's enemies was repeated in other cases; biological control, entomologists found, had definite limitations. In 1930 Howard admitted that introducing natural enemies was "infinitely more complicated than we had supposed twenty years ago, and . . . the early views of the Californians, based upon a single and very exceptional instance, were in fact nothing less than absurd."[51] Such problems, though, were precisely what the bureau could not afford. It depended on annual funding by Congress and had to show practical, effective methods as quickly as possible, As a result, biological control fell into disrepute. Only in a few places, notably in California, did it remain a serious alternative to chemicals, a subject for investigation by experiment station entomologists.[52]

The failure of other methods to meet public demands for ways to stop insects without long, expensive research, changes in farming practices, or long-term planning paved the way for chemicals. The triumph of chemical insecticides was due not just to the visible results they gave, but to their acceptance by a public and a farming community that valued, above all else, convenience, simplicity, and immediate applicability. That economic entomologists recognized this need and were prepared to meet it can be seen not only in the complaints that the entomologist was "losing sight of the insect," but in positive exhortations to use insecticides. In 1923, in an article on "The Need of Chemistry for the Student of Entomology," William Moore defended the emphasis on chemicals in terms of efficiency and public desires. The economic entomologist "justifies his work and makes his claim for federal and state funds upon his ability to make sound recommendations for the control of various insect pests. Although parasites may hold in check certain insects and damage from others may be reduced by proper cultural methods, the burden of control rests upon the use of insecticides."[53]

There were a variety of other reasons that helped make insecticides so popular—profit from sales, the new generation of entomologists trained in the land-grant colleges, and the popular idea of man and insect locked in a "battle for survival." Profit was the most tangible and probably the strongest motive. Alone among insect remedies, chemicals lent themselves to quick, large-scale commercial exploitation, and by 1910 there were so many manufacturers, and there was so much fraud, that Congress passed the Insecticide Act of 1910, which established standards and a regulatory apparatus. As early as 1912 an entomologist in California complained of the "tremendous influence the manufacturers and dealers of insecticides are exerting." Their salesmen see more farmers than the county agents can, they give the last advice before the farmer applies the product, and they can "counteract our recommendation."[54] By the 1920s the bureau was working closely with commercial interests in the field, and the partnership continued to grow.

Part of the shift toward exclusively chemical control is due to the replacement of the pioneer economic entomologists, largely self-trained or educated in the naturalist tradition of the nineteenth century. The founders of the discipline had, of necessity, received a broad, general training in biology. Most had been forced, while working in the experiment stations and land-grant colleges, to teach in various fields. Several had been outstanding taxonomic entomologists, specialists in the parasitic or predaceous genera with which they worked.[55] They tended to see insects as part of nature, and to see economic entomology as part of ecology—applied ecology. Their successors, who began to enter federal and state service in about 1900, and who had been educated in the land-grant colleges, with a more technical education, focused closely on economic entomology and were less concerned about the general study of insects or their place in biology.[56] They were agricultural scientists, not biologists interested in the application of biological knowledge. The difference was clearly expressed in 1924 by the retiring president of the AAEE in his address "Pioneering in Economic Entomology." "The use of insecticides has led to a side issue that has assumed enormous proportions. Many of the later entomologists even before thoroughly studying the life of the insect begin experimenting with insecticides for the control of the insect in question. It is not well for the entomologist to lose sight of the insect."[57]

The popular image of man and insect locked in battle also contributed to the popularity of chemicals. One of the commonest expressions in the entomological literature was the "war against insects," and popular and semipopular articles and speeches abounded with military metaphors and Darwinian (or Spencerian) notions of "survival of the fittest" and the "struggle for existence," translated into the battle of man against insect. Even Leland O. Howard, an enthusiastic advocate of biological control, entitled his autobiography *Fighting the Insects* and produced *The Insect Menace* and *The Housefly: Disease Carrier*.[58] It is impossible to prove, but difficult to resist, the idea that this constant image of battle and war influenced people to choose, or to accept, the use of the only weapons that promised "victory." The image of war and the picture of a struggle for the "survival of the fittest" implied fixed and final solutions; it denied coexistence between man and insect. Both cultural and biological methods, however, were based on just such a coexistence, man and insect living together—if not in harmony. Chemicals, on the other hand, seemed to promise "victory."

It was into this milieu that DDT was introduced in 1942, and to a group looking for better insecticides it appeared as a gift from heaven. One economic entomologist who tested DDT on potatoes described the results, thirty years later: DDT, he said, "was a miraculous insecticide. . . . We had never seen potato plants that looked as they should before [that is, without insect damage]."[59] DDT did seem to be an ideal material. It was light in weight, inexpensive, very toxic to many different insect species but low in mammalian toxicity, efficient as both stomach and contact poison, and extremely persistent. As little as one-fourth to one-fifth of a pound, dissolved in fuel oil and sprayed over an acre, killed almost all the mosquitoes, and a single spray on a wall killed insects landing there for up to six months.[60] To many it marked the beginning of a new era. Some speculated that with the use of DDT, man might banish all insect-borne diseases from the earth, and in January 1947, Clay Lyle, retiring president of the American Association of Economic Entomologists, called for even more ambitious plans. Citing the rise in appropriations for entomological programs following the introduction of DDT, Lyle called for a program to take advantage of the entomologists' new prestige and the new weapons at their disposal: an all-out effort to wipe out native as well as imported pests. "The time has now arrived for the eradication of the house fly and with it the horn fly. . . . This is not a fantastic dream but something that is almost certain to happen."[61]

It remained a fantastic dream, but even more sober plans called for the eradication of imported pests. By the mid-1950s the USDA had organized and secured funding for several large campaigns against various pests, campaigns covering millions of acres and involving enormous volumes of insecticides. Against the gypsy moth alone the USDA sprayed over 4,000,000 acres between 1954 and 1958. In 1963, the year after *Silent Spring* was published, the department was still conducting eighteen major campaigns against various insects, and planned to spray 3,200,000 acres that year. And despite a report by the President's Science Advisory Committee calling the campaigns expensive failures, Secretary of Agriculture Orville Freeman said that the USDA planned to continue eradication work.[62] It would take more than criticism to deter entomologists from following the successful, and generally acceptable, path they had followed for more than fifty years.

CHAPTER 2

Human Health and
Insecticide Residues

JUST AS THE ARSENICALS came to dominate insect control measures, so too did they provide the basis for federal and state regulation of insecticide residues as health hazards. The Department of Agriculture's Bureau of Chemistry, predecessor to the Food and Drug Administration, began setting standards for insecticide residues on food in the mid-1920s, two decades before the introduction of DDT. By the time the new chemical was available, experience with the arsenicals had generated a set of working arrangements among the parties concerned with the problem—doctors, regulatory officials, and farmers—and a set of assumptions about the nature and extent of the residue problem that determined the response to DDT. In the late 1960s new scientific information, several industrial-health problems (including asbestos and vinyl chloride), and the introduction of public interest groups into the debate began to break down the consensus. For the first two decades of DDT's use, however, arguments about DDT and human health took place in closed professional circles, a set of debates among adversaries hearkening back to the lessons learned, or not learned, in the period before DDT replaced the older insecticides.

The government's involvement with the problems of agricultural chemicals and the public goes back to the early years of insecticide use, when suspicious farmers feared that the arsenic on the potato vines might poison them. Cases of accidental poisoning of livestock only increased fears, and economic entomologists who saw in paris green a new and important method of insect control set out to convince the farmers that the poisons could safely be used. Work done at several agricultural experiment stations was vital, for men

close to local problems were often the most interested and
effective advocates of insecticides. A. J. Cook of the Michigan Agricultural Experiment Station was a particularly active
propagandist for arsenicals and conducted several dramatic,
but somewhat crude, demonstrations to bolster faith in the
product. He sprayed a fruit tree with a double strength
solution of london purple, a popular arsenical, then fed the
grass under the tree to his horse. The horse lived. A. J.
Kirkland, an entomologist with the Massachusetts Agricultural Experiment Station tried the same experiment in his
state, evidently hoping to convince farmers that protecting
their trees would not endanger their livestock. C. P. Gillette
of Iowa analyzed sprayed apples and concluded that the
residues that did not weather off were insignificant, and in
New Hampshire W. C. O'Kane in 1917 conducted similar
studies on small fruits.[1]

The campaign of reassurance, which was quite successful,
was entirely the work of entomologists, who usually had no
medical training but who did have a clear interest in proving
insecticide sprays safe. There was, however, little opposition
to their work. One of Cook's colleagues at Michigan, a
chemist named R. C. Kedzie, was almost alone in pointing to
possible dangers from small, repeated doses of insecticides.
There was little medical interest in the problem. One
physiologist who was interested was Anton J. Carlson of the
University of Chicago. Writing to O'Kane about his studies
in 1917, he summarized what became the main issue for
several decades.

> Speaking as a physiologist interested in public health I
> should say that the question is not how much of the
> poison may be ingested without producing acute or
> obvious chronic symptoms, but how completely can
> man be safeguarded against even traces of poison. There
> is no question in my mind that even in less than so-called
> toxic doses lead and arsenic have deleterious effects on
> cell protoplasm, effects that are expressed in lower
> resistance to disease, lessened efficiency, and shortening
> of life.[2]

Most chemists and entomologists who were concerned with the problems, though, compared the amount of arsenic on fruit to the amount necessary to produce illness or death and concluded that there was nothing to fear, and most physicians remained unconcerned, or unaware of, the possibility of danger from insecticide residues.[3]

The first public health scare involving residues was in 1891. In late September of that year a curious citizen brought a bunch of grapes to the offices of the New York (City) Board of Health and asked for an analysis of the bluish green deposit on the fruit. Under the headline "Poison Grapes for Sale," the *New York Times* reported that the analysis showed copper sulfate with "paris green possibly mixed in it." The next day, in a story titled "More Grapes Condemned," the *Times* said that the Board of Health had seized and condemned 258 crates of grapes. There was a public panic, much like the cranberry scare of 1959. The market for grapes collapsed and the Boston and Providence Boards of Health began checking incoming shipments of New York fruit.[4] Faced with serious losses, the grape growers appealed to the U. S. Department of Agriculture, which sent Beverley T. Galloway, a plant pathologist, to New York. Galloway made an independent analysis and a public report. The deposit, he said, was not paris green, but bordeaux mixture, a solution of copper sulfate and lime commonly used as a fungicide. There was no danger of poison. An adult would have to eat 300 pounds of grapes a day, including the heavily coated stems, to get a harmful dose. The public began to eat grapes again and the *Times,* whose stories had contributed to the situation, now began to call for an examination of the acting chemist, who, the paper hinted, had obtained his job without proper qualifications.[5]

The grape scare, though a false alarm, had all the elements farmers and agricultural officials came to fear: a report of poison, wide and sensational publicity, overreaction by the public, and the threat of serious economic damage, both immediate and long-term, to growers. It also showed what became the common reaction of the USDA, which saw itself as a service agency for agriculture. It responded to the grape

growers' appeal with expert advice and publicity to counter
the newspaper stories. The chief difficulty with the USDA's
serving as the advocate of the farmers' interests was that it
also became, by law, the defender of public health. In 1906,
when Congress passed the Pure Food Act, it placed the
enforcement of that law in Harvey W. Wiley's Bureau of
Chemistry. It seemed a natural choice; Wiley was the
strongest and most tireless advocate of such a law and the man
most closely identified with the crusade for pure food.
Unfortunately, the Bureau of Chemistry was part of the
Department of Agriculture and Wiley was subordinate to the
Secretary of Agriculture, even in the enforcement of his
treasured law.[6]

Congress did not resolve this anomaly until 1940, when it
transferred the Food and Drug Administration (FDA) to the
Federal Security Agency.[7] As a result, insecticide residue
policy was formed in a department with conflicting responsibilities and needs. When insecticide residues became a
problem in the 1920s, one agency of the department, the
Bureau of Plant Industry, was recommending spraying
schedules to kill insect pests; another agency, the Bureau of
Chemistry (replaced by the Food and Drug Administration in
1927), was seizing produce, sometimes sprayed in strict
accordance with the schedules, on the grounds that it was
contaminated with poisonous chemicals. The Secretary of
Agriculture had to satisfy the requirements and needs of public
health without alienating the main political support of
the department, the farmers. When Congress had passed the
Pure Food Act in 1906, there had been no conflict; the grape
scare had been a false alarm and there appeared to be no
hazard from insecticide residues on food. In the next two
decades, though, the situation changed drastically, particularly in the apple industry, as technological progress and
geographical changes created a potentially serious problem.
By the 1920s the center of the national apple market had
shifted from the generally well-watered area of upper New
York to semiarid valleys in central Washington, where irri-
gated orchards were necessary. At the same time improved

sprays, incorporating new binders, meant that the residues
were more resistant to weathering. With less rain to wash off
the stickier sprays, fruit with more and more residue began to
reach consumers.[8]

The first indication that there might be a problem came in
the fall of 1919 when an inspector for the Boston Board of
Health found, on a pushcart, some pears heavily coated with a
floury deposit. It proved to be lead arsenate. The Bureau of
Chemistry traced the pears to a shipment from the West
Coast and promptly began an emergency survey of the crop
from that area. It concluded that the problem was a local one,
caused by poor practices on the part of a few careless growers.
In the next few years the bureau concentrated on fruit from
the Pacific Northwest, sampling shipments, lecturing growers on the need to clean their fruit, and occasionally, holding
up a shipment for cleaning on the grounds that it was
contaminated. It did not, however, take legal action or set an
official tolerance level.[9] The informal tolerance level remained
a secret even from the growers, and, despite the demonstrated
hazards of the Boston pears, the bureau did not even warn
consumers to clean fruit before eating it. Publicity, it feared,
would only cause a public panic that would harm the fruit
industry.[10]

Hopes that this policy would solve the problem disappeared in October 1925, when, in England, an analytic
chemist attached to the staff of the Hampstead Board of
Health traced two cases of arsenic poisoning to the consumption of sprayed American apples. There was a strong reaction,
which threatened for a time to end the English market for
American fruit. The Crown prosecuted several vendors for
selling apples contaminated with arsenic, there was questions
in the House of Commons, and Sir George Buchanan, an
official of the Ministry of Health, came to the United States to
discuss the situation with American officials. After conferring
with Buchanan, the Bureau of Chemistry, in cooperation
with fruit growers, worked out a certification system for
insuring that export shipments met the British tolerance
level.[11] Chemists, paid by the shippers and packers, but

licensed by state or federal examination, analyzed each shipment before it left the country. The arrangement had no legal
standing, but the growers' interest in maintaining their market and reputation enforced it.[12]

Although the British action did not receive wide publicity
in the United States, it made a domestic tolerance level
imperative; regulating export shipments without making
some provision for safeguarding Americans would have been
impossible. The bureau, though, encountered strong resistance to the establishment of any arsenic-tolerance level for
interstate fruit shipments. At the request of western fruit
growers, Senator Charles Waterman of Colorado introduced
a bill that would have exempted fresh fruit and vegetables
from the provisions of the Pure Food Act. The measure
passed the Senate without opposition, as a rider to the Surplus
Control Act, and failed to become law only when the House
struck out the entire body of the Senate bill and passed its own
version under the same title.[13] In the summer of 1926, as the
bureau attempted to bring growers into compliance with
some standard, a delegation of western fruit growers protested to Secretary of Agriculture William Jardine that they
could not meet the British standard, the accepted international
level. Agents attempting to educate the growers to the
necessity of cleaning their produce met with little cooperation; there was talk of tar and feathers and one Oregon
orchardist, Llewellyn Banks, threatened to shoot the head of
the Bureau of Chemistry's western district. Banks finally
settled for a suit charging the bureau, state officials, and the
Southern Pacific Railroad with "conspiracy, persecution, and
malicious interference with the normal trend of business."[14]

The department recommended to farmers the same means
of reducing residues it had pressed on exporters; mechanically
washing the fruit after harvest. Developing washing equipment and manufacturing it on a fairly large scale, though, was
a long-term project—as was educating the growers—and the
bureau had to take immediate action against contaminated
fruit. It began seizing shipments late in 1925, but avoided
punitive action. It allowed shippers to clean their fruit or have

it destroyed; it did not press charges. The first case in which
the defendant contested the seizure came in the summer of
1926, when the bureau seized 3,102 cases of apples and 532
boxes of pears belonging to Llewellyn Banks. To prove that
the fruit was a health hazard, the bureau introduced analyses
showing residues of up to 0.11 gr./lb., the results of animal
experiments showing that guinea pigs became ill when fed
peels from the fruit, and the testimony of several expert
witnesses, including Carlson of Chicago and Dr. Arthur S.
Lovenhart of the University of Wisconsin, two prominent
physiologists. Banks's lawyer offered no defense. He could
only remind the jury that Banks had sprayed the fruit in
accordance with the recommendations of the USDA. It
would be unfair, he said, for the department to condemn the
fruit after approving the spraying. The judge ruled that this
argument was irrelevant; the law was concerned not with
recommended sprays but with public health.[15]

Although the jury found for the government in the Banks
case, there was still need for a legally defensible standard,
which could best be met by setting tolerance levels on the
basis of medical evidence. In January 1927, Jardine called a
conference of toxicologists and physiological chemists to
consider the problem of spray residues and public health and
to make recommendations. The precedents for such an ad hoc
committee went back at least to 1907, when President Theodore Roosevelt established the Remsen board to rule on
disputed scientific matters concerning the enforcement of the
Pure Food and Drug law.[16] This new committee was probably chosen for the same purpose as the earlier one: its
nonpartisan nature and the qualifications of the members
would help deter criticism by giving an appearance of neutrality, impartiality, and expertise to the decision. In view of
the fruit growers' strong and influential support in Congress,
Jardine needed strong scientific support. The members of the
commission, usually called the Hunt Commission after its
chairman, Dr. Reid Hunt of Harvard University, were familiar with the spray-residue problem. One of them, Dr. Carl
Alsberg, had been chief of the Bureau of Chemistry from

1913 to 1921. Two others, Dr. Arthur S. Lovenhart and Dr.
Carl Voegtlin, head of the U.S. Public Health Service, had
been expert witnesses for the government in the Banks case,
as had A. J. Carlson, who had been invited to join the
committee but had been unable to attend. The other members
were F. B. Flinn and Haven Emerson, toxicologists from
Columbia University.[17]

The committee took a much more serious view of the
dangers of residue poisoning than had the farmers or the
Bureau of Chemistry. The arsenic tolerance level, it said,
should be set at 3 parts per million (ppm) or 0.021 grains per
pound (gr./lb.), the lead tolerance level at 2 ppm (.014
gr./lb.).[18] It stressed that these should be interim tolerance
levels only, pending further research, and urged as a "matter
of fundamental economic as well as social and health importance to the food industry . . . that researches be pushed
vigorously through the resources of the Government in order
to discover a substitute for lead arsenate." To fill "gaps in
existing knowledge" the committee recommended a comprehensive study of the absorption and excretion of lead and
arsenic, their circulation through the system, and their storage
in the body. Though the evidence for widespread poisoning
from spray residues was "scanty and unconvincing . . . the
insidious character of accumulative poisoning by these substances is known to be easily overlooked [sic], and . . . the
lack of evidence of prevalence of such poisoning must not be
accepted as proof that such poisoning does not exist."[19]

Even the Hunt Commission recommendation, well above
the British arsenic tolerance level of 0.010 gr./lb., was too
radical for the Department of Agriculture. When Jardine
announced a tolerance level in February 1927, he set an arsenic
tolerance level of 0.025 gr./lb and no lead tolerance level at all.
The commission report was never published, nor was there
any explanation for the absence of a lead tolerance level. The
agency believed the level had to be set by what could be
achieved, not by what was desirable. The arsenic standard,
one official explained, was based on "the department's
knowledge of what the industry could do with the most

earnest efforts at cleaning." The level would be a "fair tolerance until better equipment was perfected for washing."[20] The new Food and Drug Administration, formed in 1927 from the old Bureau of Chemistry, then lowered the amount of arsenic allowed, year by year, until in 1932 it reached the British level.[21]

The setting of a lead tolerance level illustrates the scientific difficulties in the way of enforcement. The chief obstacle was the lack of a suitable analytical test for lead. There were tests available, but they required more than a day. This posed no problem for the analyst—who could let his solutions sit while he did other work—but it was an insurmountable difficulty for the bureau. By the time it had the analytical results from a carload of fruit, the cargo would already have been shipped, broken up into small lots, and perhaps, sold. Testing before the lots were assembled for shipment, on the other hand, was completely impractical in terms of money, manpower, and time. Until 1932, when the FDA's chemists developed a quick and reliable test for lead, they assumed that the hydrochloric acid washes then in use would remove both lead and arsenic equally well, and they relied on the arsenic test as a measure of the total residue.[22] On 21 February 1933 Secretary of Agriculture David F. Houston set a lead tolerance level of 0.025 gr./lb., almost twice that recommended by the Hunt Commission and well above the level the FDA had assumed would be on fruit meeting the arsenic tolerance.

Scientific problems, though, were minor compared to political ones. Attempts to lower residue levels met with strong resistance from fruit growers, and, aided by their Congressional allies, they were able to fight off attempts to tighten standards during the New Deal years. Rexford G. Tugwell, who became Assistant Secretary of Agriculture in the first Roosevelt administration, soon found that there were severe limits to his power in this respect. When, in April 1933, Tugwell found that the Department of Agriculture was allowing residues of lead and arsenic on fruit, he was shocked. After a talk with Walter G. Campbell, head of the FDA, he admitted the impracticality of a zero tolerance level on lead arsenate, but took vigorous action to tighten standards. On 2 April he set the lead tolerance level at the Hunt Commission's figure (0.014 gr./lb.) and stated that a zero tolerance level ws desirable. He then tackled the problems posed by the fluorine compounds then coming on the market, declaring that the extreme toxicity of this element made necessary a zero tolerance level, effective immediately.[23]

The International Apple Association, the apple growers' lobby, protested vigorously. Citing Tugwell's lack of consultation with the industry and the enormous amount of fruit already certified and stored from the previous season, the association said that the new tolerance was unfair and that it would be impossible for the growers to meet the stringent new tolerances during the coming season. On June 20, Secretary of Agriculture Henry A. Wallace bowed to this pressure. He raised the lead tolerance level to 0.020 gr./lb. and established a tolerance level of 0.010 gr./lb. on florine. He reiterated the desirability of a zero tolerance level on lead and forecast a 0.014 gr./lb. tolerance level for 1934, but noted that the replacement of lead arsenate would have to wait on the development of new insecticides. The next year he repeated his forecast and his hope, but nothing came of it.[24] Residue tolerance levels remained essentially the same throughout the 1930s; the lead tolerance level was never set below 0.018 gr./lb. (1935–1937), and it was raised again to 0.025 gr./lb. in 1938.[25]

Besides deferring to the growers' desire for more time to meet standards, the FDA was careful to avoid publicity. In 1936 W. C. Geagley, a chemist with Michigan Department of Agriculture, complained that the public was not fully aware of the dangers of spray residues and charged that the authorities were afraid of a public reaction that would harm the fruit industry.[26] Paul Dunbar, then assistant commissioner of the FDA, admitted this in an article written in 1959. Our "objective," he said, "was to persuade all departmental agencies to cooperate in working out the [spray residue] problem and to refrain meanwhile from creating public alarm."[27] Swann Harding, a Department of Agriculture information officer who wrote articles urging drug law reform, managed, when he wrote about spray residues, to stress the harmlessness of the small amounts of residue the department allowed on food. Even Ruth Lamb, whose book, *American Chamber of Horrors,* aroused a storm of opposition from the drug industry, glossed over several important points in her discussion of spray residues, including the specific recommendations of the Hunt Commission and the matter of a lead tolerance level.[28]

The Hunt Commission had strongly recommended toxicological studies to establish the chronic effects of residues, but the government did little to meet this need. The FDA did not get appropriations for a continuing analysis of foods in the normal diet until 1934, or appropriations for experimental studies over the lifetime of animals until 1935.[29] Other investigators did, however, conduct studies, and their results made it even more difficult to defend the FDA's policies. Between 1929 and 1933 medical researchers at the Stuyvesant Square Hospital in New York City published several papers that indicated the residue situation was more serious than the FDA admitted.[30] Their work included reviews of the literature, extensive tests on food, clinical observation of cases of lead and arsenic in many common foods, and several cases of extreme carelessness in the use of sprays, including the spraying of cattle feed with arsenicals. "[T]he present spray residues situation," they concluded, "constitutes a menace to the general well-being of individuals."[31] In 1935 and again in 1937 the American Medical Association took the position that the residues might become a public health hazard of the first order, and it urged further study and more stringent state regulation of the use of dangerous insecticides.[32]

Despite the evident medical concern, there was little that the FDA could do to lower tolerances. It was the middle of the Great Depression, and the growers, forced to operate on a smaller profit margin and conscious of every production cost, conducted a strong campaign against federal regulation. Besides direct complaints to the Department of Agriculture and to the President, they took their case to Congress, where they found sympathetic listeners. One of the these was Clarence Cannon of Missouri, a member of the Subcommittee on Agricultural Appropriations of the House Appropriations Committee. Cannon did not believe that spray residues were dangerous; "[L]ead arsenate on apples never harmed a man, woman, or child" he often remarked.[33]

The FDA's troubles with Cannon show the extent of Congressional pressure and the dependence of even well-established agencies on interest groups and strategically placed Congressmen. The episodes are not remarkable; they are, unfortunately, only too typical of the problems of "independent" agencies. In 1935, at the hearings on the Department of Agriculture budget for the coming year, Cannon questioned the bureau chiefs from the department about the toxicity of lead arsenate, asking particularly if they knew of any cases of actual poisoning from spray residues. They all replied that the matter lay outside their jurisdiction; the man to ask was Walter G. Campbell, commissioner of the FDA. When Campbell appeared to testify on 11 February, Cannon asked him if he had any evidence on the toxicity of lead arsenate sprays. Campbell did. He spoke for an hour, citing several cases of poisoning, describing the deaths of two children, and introducing reports from physicians and autopsy results on these cases. He stressed, however, that the chief danger was not acute poisoning, but the cumulative deterioration of liver, kidneys, and other organs by the continued consumption of small amounts of lead and arsenic.

When the record was printed, however, none of Campbell's testimony on residues appeared, leaving the impression, in the face of Cannon's questions to the other bureau chiefs, that the FDA had no evidence that spray residues were harmful or that there was any reason for the tolerance levels.[34] Fortunately for the FDA, Senator Royal Copeland of New York, who was sponsoring the agency's bill to revise the Pure Food Act, "found out" about the omission. With a copy of the official record he pointed out the "error" and threatened to read Campbell's testimony in full on the Senate floor. Though the printer got the blame for the incident, it seems that one of the committee members, presumably Cannon, had ordered the deletion. Following Copeland's speech, James F. Buchanan of Texas, chairman of the House Appropriations Committee, questioned Campbell about the incident and promised that in the future all his words would be printed.[35]

The growers did not rely completely on Cannon. Their own lobby worked hard to modify various provisions of the FDA bill (which eventually became the 1938 Food, Drug, and Cosmetic Act). Samuel Fraser, secretary of the International Apple Association, testified during Senate hearings on two versions of this act, S.2800 (1934) and S.5 (1935),[36] contending that the provisions giving the Secretary of Agriculture power to set legally binding tolerance levels were arbitrary exercises of power, dangerous to liberty, and entirely contrary to the principles of Anglo-Saxon law. They transferred the burden of proof from the plaintiff, the government, to the defendant, the grower, and left him with no recourse if he wished to challenge the rulings of the Secretary, however arbitrary and capricious they might be. The International Apple Association wanted the full range of the Secretary's decisions to be subject to the same legal review provided by the 1906 law. In 1934 Fraser sounded the refrain that ran through his testimony when he asked that "the Government should be required to prove its case and to prove by competent evidence that the amount alleged to be found is or may be injurious to health. The defendant should be allowed to have his 'day in court.' "[37] This, of course, was the time-consuming and repetitive process the FDA wished to avoid by amending the law. It wanted to prove a standard once and give it the force of law, not be compelled to prove the same case every time it went to court.

The growers stuck to their position throughout the long fight over the revision of the food and drug laws in the 1930s and earned the distinction of holding the last ditch against passage of the Food, Drug, and Cosmetic Act of 1938. Their objections were finally met by allowing appeal of the Secretary's tolerance levels to the ten U. S. circuit courts.[38] Thus,

twenty years after the first spray-residue investigations, the
FDA got statutory authority to set tolerance levels for residues on food.

The 1938 law simplified enforcement of spray-residue
regulations but the agency was less successful in making the
studies that would have provided a scientific basis for its
decisions. Its opposition was, again, Cannon. In 1935 it had
started a project to determine the harmful effects of residues, a
study involving the observation, over their lifetime, of experimental animals fed small amounts of lead and arsenic. In
1937, however, when Congressman Buchanan died, Cannon
succeeded to the chairmanship of the agricultural appropriations subcommittee. He lost no time in asserting his authority; the appropriations act for fiscal 1937 contained the
provision "that no part of the funds appropriated by this act
shall be used for laboratory investigations to determine the
possible harmful effects on human beings of spray insecticides
on fruits and vegetables." This stopped the FDA's program,
and paved the way for an appropriation of $50,000 annually to
the U. S. Public Health Service "for investigations to determine the possible harmful effects on human beings of spray
insecticides on fruits and vegetables."[39]

The Public Health Service had a different approach to the
question of the effects of residues than did the FDA. The latter
had set up an experimental study to determine the effects of
lead and arsenic on physiological functions, using experimental animals exposed over a lifetime. The Public Health
Service's study was neither experimental nor designed to
study the chronic effects of lead and arsenic. It was a survey of
a population, orchardists and consumers, living in an apple
growing area and subject to greater exposure than was the
normal population. Tests included clinical and laboratory
examinations, and the program was designed to discover if
characteristic forms of ill health resulted from ingestion of
lead arsenate spray, to find out if occupational exposure
produced any characteristic symptoms, to obtain analytical
data on lead in the body and in bodily wastes, and to find if
there was any relationship between lead arsenate spray and
susceptibility to other diseases.[40]

The Public Health Service chose the river valley around Wenatchee, Washington, an area that raised 20 to 25 percent of the country's apple crop and used up to 4,500,000 pounds of lead arsenate annually. A three-year study, consisting of laboratory and clinical studies of 1,231 persons, including orchardists, part-time orchard workers, the normal adult population, and children, revealed that only seven people, six men and a woman, "had a *combination* of clinical and laboratory findings referable to the absorption of lead arsenate." Even these cases were not up to the medical criteria for lead poisoning.[41] Although many of the subjects, including children, had higher levels of lead and arsenic in their blood than the normal population, there was no pathology that could be related to lead arsenate, or any diseases traceable to the high level of residues.

The most important conclusion of the study—that cases of lead and arsenic poisoning were rare and not clinically important—shows the primary difference between the Public Health Service's approach to the problem and that of the FDA. As early as Carlson's letter to O'Kane in 1917, medical scientists had warned that the major problem in studying the health hazards posed by residues was the vague and ill-defined nature of the symptoms. The Hunt Commission had reiterated this position in 1927, pointing out that symptoms were easily overlooked precisely because they were not those of classical poisoning. The situation, though, called for intense effort on the part of physicians and medical scientists, for if residues did have chronic effects, in the doses the public was receiving, then a large part of the population would, ultimately, be affected. The FDA study had been designed to yield such data for experimental animals, data that could be used for an extrapolation to humans. The Public Health Service had ignored the problem. By using the criteria for acute poisoning, the very thing that no one claimed was a serious problem, it defined the real question—the effects of low-level doses over a long period—out of existence.

As a result of the Public Health Service's recommendations, made on the basis of the Wentachee study, the Federal Security Administrator (the FDA had been transferred to the

Federal Security Adminstration, predecessor of the Department of Health, Education, and Welfare [HEW], on 1 July
1940) changed the tolerance levels of lead and arsenic on fresh
fruit on 10 August 1940. The new arsenic tolerance level was
0.025 gr./lb.[42] There was no argument that public health
would be improved by the higher levels, only that the public
apparently could stand them and the farmer profit from them,
spraying with less caution and cutting the expense of cleaning
his crop. It was the last change made in the tolerance levels
before DDT replaced lead arsenate in orchard sprays.

The controversy over spray residues in this period, which
set policy for the early period of DDT use, revolved around
the question of safety: how best could medical scientists
establish the safety, or harmful effects, of low levels of
residues on food? The two groups in the battle differed in
quite fundamental ways. One side—comprised of the
farmers, their Congressional allies, economic entomologists,
the Public Health Service, and a respectable number of
medical scientists—believed that the criteria for safety were
those of clinical poisoning; doses that did not produce illness
were safe. The best way to ascertain safety, they thought, was
to study groups of people exposed to higher-than-normal
concentrations of residues; if these people were healthy surely
the public would be too. Finally, they insisted, the burden of
proof should rest on those who would ban a chemical from
use. This was the rationale behind the Public Health Service's
work and the argument advanced by farmers and economic
entomologists. The other group—the Food and Drug Administration, consumer groups, and some medical scientists and
physicians—took an entirely different view. Safety, they
thought, should not be determined by signs of classical
poisoning, but by the evidence of chronic effects, and such
data could best be obtained by extrapolating from laboratory
tests, preferably conducted on experimental animals over
their lifetime. Residue standards should not be set by the
amount the public could apparently tolerate, but by the
evidence of chronic effects, and they should safeguard the
public as much as possible from danger. The burden of proof

should rest on those who used the chemicals and exposed the public to danger.

That the fight was one-sided is clear evidence of the political power of the growers and of the public's lack of interest in the problem. The revision of the food and drug laws in the 1930s, during which the two factions in the residue battle attempted to have their ideas written into the law, attracted little interest. Even the rise of organized pressure groups for consumer interests in the 1930s did not significantly affect the orientation of the 1938 Food, Drug, and Cosmetic Act. It was, to be sure, different from the 1906 act in that it is written and supported by an agency of the government, rather than having the support of individuals and private groups, but the act was passed neither by nor for the public. David Cavers, a law professor who, in 1933, helped draft the first version of what became the 1938 act, wrote in 1939 that "perhaps the most striking characteristic of the history of the Food, Drug, and Cosmetic Act is the fact that this measure . . . never became the object of widespread public attention, much less of informed public support."[43]

By the time DDT became available for civilian agricultural and public health use in 1946, both the Food and Drug Administration and the Department of Agriculture were accustomed to acting as service agencies for interest groups. Their concern, except where there was an immediate danger to public health or safety, was to work with the affected industry and to enforce compliance without causing the industry more difficulty than necessary—the industry being the judge of the hardship. Whether, under the circumstances, they could have followed different policies is another question, but given their need for annual funding, the lack of broad public support for a strict policy, and the imprecise nature of the threat to public health, it is difficult to see how else they could have acted. From the viewpoint of the 1970s the policy may seem less than wise, but largely because of our increased knowledge.

Part Two

LEARNING ABOUT DDT

CHAPTER 3

Applying Old Lessons

WHEN, in the late 1960s and early 1970s, states and the federal government began to restrict or ban the use of DDT, they did so because of the chemical's effects on the environment. Although there was some concern about its alleged carcinogenicity, human health was not a major issue nor a major part of the scientific case against DDT. In this respect DDT was quite different from earlier insecticides, which had been applied to agricultural areas and had their effects on human consumers. DDT was widely used in forests, swamps, and other habitats where human involvement had been minimal. Its effects on the environment, however, were believed to be worldwide. The debate over DDT did not begin in this context, though. Almost all the early scientific discussion of the side effects of the new chemical centered on its impact on the human body, as medical scientists and regulatory officials attempted to fit it into the mold provided by experience with earlier insecticides. Between 1943, when the wartime tests began, and the mid-1950s, when a modus vivendi was worked out, there was a replay of the struggle over regulatory policy from the 1930s, a struggle in which, though some of the actors were different, the lines were much the same. The official debate over regulatory policy for DDT was, however, conditioned by a factor unique in the history of insecticide regulation: DDT was first used during World War II; by the time it entered the civilian market it already had a reputation for effectiveness, power, and safety unmatched by any other material.

When the United States entered a two-front war in December 1941, it confronted several serious medical problems. One of the first priorities of the new Committee on Medical Research (of the Office of Scientific Research and Develop-

ment) was the protection of American troops against insect-
borne disease. Troops fighting in Europe or Africa would be
exposed to typhus, which throughout European history had
proved as potent a killer as arrows and bullets. Soldiers in the
South Pacific would need protection against malaria and other
tropical diseases carried by mosquitoes. Nor were troops the
only concern. Dislocated civilian populations also would
require protection, lest the Allied armies find disease spreading behind their lines. Unfortunately, there seemed little hope
of doing these jobs with available materials. Botanical insecticides were ideal for louse powders, but the plants were in
short supply and were grown in areas such as Africa, the
Dutch East Indies, and South America, which were inaccessible or required scarce shipping, often through enemy-
controlled waters. Against malarial mosquitoes there was not
even this hope. No known insecticide would provide effective
protection for forces on South Pacific islands.[1]

In mid-1942 the solution appeared; the American representatives of the Swiss firm of J. R. Geigy Company brought to
the Department of Agriculture some insecticide samples they
had received from the Swiss office—100 pounds each of a 5
percent spray and a 5 percent powder, trade-named
"Gesarol." H. L. Haller of the Division of Insecticide Investigations (part of the Bureau of Entomology and Plant Quarantine [BEPQ]) analyzed it and worked out a synthesis while the
BEPQ and the War Food Administration tested insecticidal
efficiency of the new material. The active ingredient, Haller
found, was 1,1,1 trichloro-2,2bis (parachlorophenyl)ethane,
commonly called dichloro-diphenyl-trichloro-ethane (the
first letters of the parts of the common name are DDT, hence
the standard abbreviation). Othmar Ziedler, a German
chemist, had synthesized the compound in 1874 as part of his
work on the substitution products of aromatic hydrocarbon
compounds, but not until the 1930s, when Paul Muller's
research group at Geigy tried it against potato beetles and
clothes moths, had anyone found a practical use for the
chemical.[2]

Muller's group had found DDT's insecticidal properties "wonderful," and American tests conducted in 1942 and 1943 confirmed this judgment. What, though, of its effects on man? Could troops stand the intense, if limited, exposure that would be involved in mosquito and louse control campaigns? The Committee on Medical Research arranged for an investigation, calling in experts from the National Institutes of Health, the Food and Drug Administration, the Kettering Laboratory of the College of Medicine of the University of Cincinnati, the National Research Council, and the Public Health Service, as well as representatives of the armed forces' medical corps. Although the program they devised included tests of chronic toxicity and of persistence and accumulation in the body, it emphasized immediate safety to operators of spray equipment and directly exposed persons.[3]

The results were very encouraging—DDT's acute toxicity to humans was very low—and by the middle of 1944 the armed forces had made DDT standard issue. The result was a massive outpouring of money and a large investment of time by scientists and engineers to provide the full range of equipment for producing and using DDT. The BEPQ developed spraying equipment, from aerosol bombs to tanks and nozzles for aerial spraying. The government had already assigned contracts to several chemical companies to produce DDT, and it quickly moved to expand production. By the end of 1944 American manufacturers were turning out over 2,000,000 pounds per month of DDT for military use. By the end of the war production was about 3,000,000 pounds per month.[4] When DDT entered the civilian market there was available a complete range of surplus equipment, techniques for using the new chemical, and an enormous public demand. "The publicity given DDT," a food company official later commented, "might well be envied by any Hollywood movie star."[5]

One of the first field uses for DDT was in the louse powders that halted the Naples typhus epidemic in 1943–1944, and it was here that DDT gained its reputation as a "miracle" insecticide. As early as July 1943, isolated cases of typhus had been reported in the city, but with the coming of winter and confinement of the population indoors, the disease assumed epidemic proportions. American and British occupation authorities began dusting exposed persons to kill body lice, using pyrethrum (a botanical insecticide) and a new dusting method developed by the Rockefeller Foundation Health Commission and the U.S. Typhus Commission. A squirt gun forced powder into the subject's sleeves, waistband, collar, pants' cuffs, hair and hat, a technique simple enough to permit dusting everyone in contact with a typhus victim. It replaced a cumbersome system that had involved baking the subject's clothes, shaving head and body hair, and painting the shaved areas with an ovicide. The case-reporting system allowed public health officers to find and dust each victim's family, fellow workers, and neighbors, killing the lice and breaking the chain of infection. By the time the epidemic was over, in February, 1944, the medical men had administered over 3,000,000 individual dustings to the citizens and the occupying troops.[6]

It was a dramatic story—the first time a typhus epidemic had been arrested by public health measures—and DDT received the credit. Newspaper accounts extolling the wonders of the new chemical found support in speculations that DDT would revolutionize public health; Surgeon General of the Army Major General Norman J. Kirk thought that DDT might mean the end of insect-borne diseases.[7] In fact, DDT had become available in quantity only after the older insecticides had broken the back of the epidemic, and the men in charge of the program emphasized the central case-reporting system and the new dusting technique. Even the assistant chief of the BEPQ, Fred C. Bishopp, who later ardently defended DDT, attributed the success of the campaign in Naples to "effective organization and energetic application of control measures."[8]

Malaria control was even more important than louse control, for malaria is endemic throughout much of the tropics. Here again wartime programs proved important in enhancing

DDT's reputation and in providing a body of equipment and knowledge for peace-time use. There were some malaria control programs in Italy, but the most spectacular were those in the South Pacific, where the military employed converted transport planes or bombers to blanket entire islands with DDT sprays, including areas that were inaccessible to ground equipment. DDT was particularly well adapted to aerial spray progams. Under ideal conditions as little as one-fourth to one-fifth of a pound per acre controlled adult mosquitoes, and, at the normal dilution rate of one pound of DDT per gallon of solution, one aircraft could easily treat a large area. DDT could easily be shipped as a concentrated solution and diluted for use with readily available solvents—kerosene or fuel oil—and its solubility prevented the clogged pipes and nozzles common with suspensions and eliminated the equipment needed to disperse insoluble materials.[9] The sprays drastically reduced the incidence of malaria, and despite heavy exposure, there were no verifiable symptoms of poisoning or illnesses among the troops.[10]

As soon as production exceeded military requirements the War Production Board allowed the surplus to be used for experiments; it released DDT for general civilian use on 1 August 1945.[11] The entry of the new insecticide posed a problem for the FDA. Although DDT's acute toxicity to humans was low, early experiments indicated that it could accumulate in the body and be passed on in milk. As early as 1944 officials were expressing, privately, reservations about using DDT until complete data on its chronic toxicity were available. It was one thing to dust or spray people in obvious and serious danger from insect-borne disease; it was quite another to dose food that would be consumed by an entire population not at risk.[12] Still, there was no way for the FDA to hold DDT off the market. The 1938 law only gave the agency the power to set binding tolerance levels after a long series of hearings. It could set interim tolerance levels at once, but it could hardly forbid the use of DDT without some good reason that would command public backing, and in the immediate postwar period there was no sentiment for such a ban or even for elementary precautions.[13] The FDA set provisional tolerance levels for residues of DDT on food products, pending further tests and public hearings to establish a formal tolerance level.[14] Since there were effective, safe insecticides for the common vegetable crops, it "did not recommend" the use of DDT. On fruit, where lead arsenate had been the most common insecticide, the FDA set an "action level," a level above which it would prosecute growers, of 7 ppm, the same level set for lead residues and for the fluorine insecticides in 1940. Because of the importance of milk in the diet of infants and invalids, the agency set a "zero tolerance" for DDT in milk and warned against using DDT on feed or forage crops for dairy cattle.[15]

The tolerance levels were not simply a routine precaution. Wartime tests had shown that high chronic exposure levels in rats caused "fatty degeneration of the liver and kidney."[16] DDT was a central nervous system poison, and persons subjected to large doses suffered from tremors, aching joints, nervousness, and depression. Recovery was often quite slow. DDT also accumulated in the body, concentrating in the fat deposits, and would be passed on in the milk.[17] Although the new chemical seemed a wonderful substitute for lead arsenate, it was not without its dangers.

Postwar tests confirmed these findings. The final report of the Committee on Medical Research, published in 1948, summarized the medical evidence on DDT's accumulation in the body, its excretion in the milk, and the progressive damage caused by chronic doses. The committee stressed the still unknown consequences of long-term exposure. "Information on the effects of the ingestion of small amounts of DDT over long periods of time is still incomplete," it warned. "This information is important since it involves the use of DDT on crops and its effects on farm animals, including milk cows. . . . This problem will require careful observation over a period of years for elucidation."[18] The committee was not alone in its belief that lack of data on chronic toxicity left a dangerous gap in public health officials' knowledge. In 1945 George Decker, head of the Economic

Entomology Section of the Illinois Natural History Survey, noted that the agricultural use of DDT raised the possibility of a hazard from the cumulative effects of eating fruits and vegetables sprayed with DDT and from meat and milk contaminated with residues.[19] Three years later the American Medical Association's Council on Foods and Nutrition warned of an "appalling lack of factual data concerning the effect of these substances [pesticides] when ingested with food. The chronic toxicity to man of most of the newer insecticides is entirely unexplored."[20]

The theoretical dangers of DDT, though, were so distant, and its superiority over lead arsenate and its effectiveness in so many situations so obvious, that it quickly came to be used in agricultural and public health insect control programs throughout the country. By 1950 DDT production was well above the peak lead arsenate had achieved in the mid-1930s, and it continued to rise. The Public Health Service used enormous amounts of DDT in mosquito control programs, and towns and cities used more to kill mosquitoes and the bark beetles that carried Dutch elm disease. In desperation, some towns even sprayed DDT in an effort to combat polio. Foresters quickly began to use DDT for aerial sprays against important forest pests, and farmers tried DDT against almost anything that flew, crawled, or walked.[21]

There was no evidence that the growing use of DDT was harming people, but there were some disturbing indications that the chemical was far more persistent and mobile than had been suspected. By 1950 there was enough evidence to make the subject of DDT residues a matter of scientific, if not public, concern. In the late 1940s, for example, scientists found that DDT applied to cow barns, even when cows were absent, soon showed up in the milk, and the FDA and USDA quickly issued a joint statement warning farmers not to use the chemical around their cows.[22] Edwin Laug and his coworkers in the FDA's Division of Pharmacology found in 1950 that DDT was showing up in the fat of persons who were not occupationally exposed, and that residues were not limited to a few cases. Seventy-five samples from the general population showed an average of 5.3 ppm DDT; twenty-four milk samples from nursing mothers had an average of 0.13 ppm DDT. Laug estimated that, after only five years of DDT use, the normal American diet had an average of 0.05 ppm DDT.[23] At the same time the Beech-Nut Packing Company, which specialized in prepared baby foods, was finding great difficulty in obtaining sufficient amounts of residue-free vegetables for its products.[24]

In 1950 and 1951 the first public debate on the safety of DDT took place when the House Select Committee to Investigate the Use of Chemicals in Food Products held hearings on the safety of food additives and other chemicals and on the legislation regulating their use.[25] The committee, usually called the Delaney Committee after its chairman, James J. Delaney of New York, investigated the entire range of additives and processing practices, including the use of chemicals in cosmetics and fluorides in drinking water. It heard testimony from representatives of agriculture, chemical and food companies, cosmetic firms, consumer groups, and assorted individuals, ranging from eminent scientists to unknown doctors. The safety of DDT residues was only a minor issue for the committee, but an important one for those concerned with agricultural chemicals, for the committee's recommendations could drastically affect the marketing, testing, or use of important and profitable materials. The hearings brought out the Public Health Service, the Food and Drug Administration, the Department of Agriculture, farm organizations, the industry's lobbies (for the first time), and physicians and physiologists concerned about public health.

Witnesses sharply disagreed on the adequacy of the laws controlling the use of insecticides and on the possible dangers of DDT residues. Insecticide manufacturers, farmers, agricultural scientists, and USDA officials thought existing legislation was sufficient to protect the public and they regarded the safety of DDT as proven. Scientists from universities, private foundations, and the FDA came before the committee to say more research was needed on the chronic toxicity of the new insecticides and more controls on their introduction and use.

There was insufficient evidence, they said, to show that DDT was safe for general, prolonged use, and current regulations were unable to prevent the growth of a serious public health hazard if the new insecticides had unsuspected properties.

The FDA made a strong case for changes in the law. Commissioner Paul Dunbar said that an insecticide could be sold before its safety was established and without identifying its active ingredient. It might be marketed without the FDA's having any idea of its effects on the human body or even of the proper analytical procedures for its detection in foods.[26] The use of DDT during World War II, he said, had been "a reasonably calculated military risk," but, he implied, other standards should govern the use of the chemical in general civilian use. There was here no question of major diseases, and exposure would not be, as it had been during the war, for a short period of time.[27] Dr. Arnold Lehman, director of the Division of Pharmacology (of the FDA) testified on the results of the FDA's latest tests. They had shown, he said, possible chronic dangers to humans from DDT residues in their food. Rats, for example, showed damage to liver function when the level in their diet was as low as 5 ppm. Directly contradicting the position of the insecticide manufacturers and of the Public Health Service, Lehman said that the potential hazards of DDT had been underestimated.[28]

Other scientists agreed that more research was needed. Dr. Robert Kehoe, director of the Kettering Laboratory of the School of Medicine of the University of Cincinnati, said that the presence of residues in virtually the entire population, although no cause for alarm, was cause for concern. Dr. Wilhelm C. Heuper of the National Research Council thought it possible that DDT might cause cancer. Although there was no record of a human cancer induced by insecticide residues, he said, the FDA tests on rats indicated that the chemical might be harmful to man, and he urged more research and monitoring of residues in the population. John Dendy, an analytical chemist with the Texas Research Foundation said that there was continuing and increasing contamination of crops. Despite the FDA warnings and the establishment of "no tolerance" on DDT in milk, he had found DDT residues in milk from Texas, Wisconsin, Missouri, and Oklahoma. This, he said, was "prima facie evidence that the existing laws and/or enforcement procedures are insufficient to prevent the development of this most serious hazard to human health."[29]

Some of the most serious complaints about the contamination of food by insecticide residues came from officials of the Beech-Nut Packing Company. L. G. Cox, head of technical projects, said that as early as 1947 the company had been forced to undertake an extensive, and costly, project to find analytical methods that would detect residues of the new chemicals. W. A. Brittin, head of the firm's food laboratory, complained that the new insecticides had been marketed "before adequate information was available on the acute or chronic toxicity of the chemicals involved." At the same time, he added, there were no "regulations limiting the use of the organic spray residues on fruits and vegetables or any information as to what levels were safe for human consumption." The new insecticides posed other problems. Beech-Nut, Cox said, had found it more and more difficult to find the necessary quantities of residue-free vegetables for its products and was now, reluctantly, accepting a finite residue level in its baby foods.[30]

Representatives of farmers and manufacturers, on the other hand, thought that there was no need for changes in legislation or cause for alarm about the safety of the new insecticides. Lea S. Hitchner, president of the National Agricultural Chemicals Association (NACA), said that "[e]xisting legislation makes possible adequate protection to the public. . . . I know of no other industry which has to comply with more laws and regulations in order to sell its products."[31] Government agencies had adequate legal authority and the full cooperation of both users and manufacturers. As an example of the effectiveness of legislation and the current system of voluntary cooperation, he said that the industry stopped sales of DDT for fly sprays in dairy barns as soon as the FDA and USDA warned against the practice. Under cross-examination by a skeptical committee counsel, though, he admitted that the industry had had no choice in the matter.[32]

The USDA and economic entomologists were sympathetic to the industry's position. Fred C. Bishopp and Edward F. Knipling of the BEPQ defended industry studies of new pesticides and manufacturers' efforts to comply with government regulations. Research, Bishopp pointed out, could go on forever without disclosing all possible hazards from expected uses. The government had to take a reasonable approach, acting on the available information while being prepared to change when the situation warranted it. Other economic entomologists took much the same position. George C. Decker, who had earlier worried about the chronic problems of DDT ingestion, now reversed himself. The present regulations, he told the committee, give a "large and perhaps adequate degree of protection from the serious hazards of food contamination by pesticides." Asked about DDT in milk, including residues that ranged up to 13.8 ppm, he said that he did not "consider it sufficiently objectionable to completely challenge the use of DDT or any other equivalent pesticidal chemical for national use."[33]

The fruit growers strongly opposed more control of pesticides and pesticide residues, particularly if the control was to be exercised by the FDA. Samuel Fraser, secretary of the International Apple Association, characterized the agitation for new legislation as a "grab for power which is to be secured under the whip of hysteria."[34] C. E. Chase, representing Washington orchardists, complained that the FDA was insensitive to the growers' economic problems, too conservative in setting residue levels, and reluctant to admit its mistakes. Its decisions on lead and arsenic residues in the 1930s, he said, had been "very costly to our industry."[35] He urged that authority for the scientific study of pesticide residues and their relation to human health be given to the Public Health Service.

There was impressive medical support for the industry's position on residues, the testimony of Dr. Wayland J. Hayes,

Jr., a U.S. Public Health Service toxicologist who had done the only extensive studies of DDT in man. Hayes believed that DDT was safe and that its safety had been established before it was released for agricultural and public health programs. The "use of DDT by the general public," he said, "was permitted only after a large amount of careful scientifically controlled study convinced the governmental authorities that DDT could be used safely under the conditions prescribed."[36] He discounted the evidence of widespread contamination of food. Citing Public Health Service studies of residue levels in human fat, he suggested that present storage levels were "the result of incidental or occupational exposure through inhalation or contact in agricultural areas." In response to committee counsel's questions about the FDA's animal experiments, Hayes pointed out that the liver damage in rats was reversible, and that the observed changes in the liver did not seem to affect the rats' health. He admitted that there was no evidence on the level of DDT below which adverse long-term effects were absent, but he did not regard this as an important point. What was important was that there were no pathological effects among users, in particular no deaths or injuries among applicators.[37]

Hayes's testimony was important not simply because he provided support for standards based on clinical symptoms of poisoning, but because he became the mainstay of the medical defense of DDT. For the next twenty years he defended the safety of DDT and the standards that the farmers and manufacturers appealed to. At hearings throughout the country, including both the Wisconsin hearing of 1968–1969 and the Environmental Protection Agency hearing of 1971–1972, he was a key witness for the defenders of DDT. His assumptions were those the Public Health Service had relied on in its prewar study of lead and arsenic residues: that the standard of safety was the lack of obvious pathology in a group exposed to higher than normal concentrations of the chemical. Hayes relied on the wartime tests, done to see if soldiers could stand the exposure necessary for mosquito and louse control, and on the exposure of millions to louse powder in occupied cities, prisons, and relocation centers. There had been, as he stressed, no noticeable illness from this or subsequent exposure.

Despite a vigorous defense of the status quo by the insecticide industry, the committee did conclude that there was need for new legislation on insecticides. Representative Arthur L. Miller of Nebraska incorporated the committee recommendations into a bill that became known as the Miller Amendment (to the 1938 Food, Drug, and Cosmetic Act). It provided for the registration of pesticides before they could be sold, and it set up a simplified procedure for administering regulations. The most important provision of the bill—to which the pesticide manufacturers strongly objected—placed on the applicant the burden of proof of safety. In order to register a chemical for use on food crops, the manufacturer had to present data showing that the expected residues for each use were too low to pose a danger to human health.[38]

The manufacturers objected on the grounds that the bill vested excessive power in the FDA, giving it authority arbitrarily to ban the use of agricultural chemicals. The amendment also left them subject to regulation by two agencies, the FDA and the USDA, and would stifle research by substituting the judgment of a single bureaucrat for the informed consensus of hundreds of research workers. They expressed fears that the government would select one material as the safest and ban all the others, depriving the farmer of the wide selection they thought he needed for effective insect control.[39] W. W. Sutherland's speech to the American Association of Economic Entomologists in December 1951 is a good example of industry attitudes. Sutherland, an executive of Dow Chemical Company, repeated Hitchner's charge that the pesticide industry was "one of the most, if not the most, highly regulated industries in the country. . . ." The Miller Amendment, he said, would give "certain government agencies arbitrary powers over the marketing of products. . . ." The agencies could require endless tests to assure the safety of a product, but "[r]esearch can go on forever without establishing all the factors which practical use of a product may eventually show up. A reasonable approach must be taken if progress is not to be seriously impeded."[40]

At the same time that Sutherland gave his speech, Dr. Frank Princi, who had testified at the Delaney hearings, addressed a joint conference of the American Phytopathological Society and the American Association of Economic Entomologists, contending that DDT was safe for general use. There was, he insisted, no clear evidence of poisoning from DDT residues, no cases of poisoning in users and no symptoms among workers in DDT plants.[41] A few months later Dr. John Foulger, director of the Industrial Safety Laboratory of Dupont and Company, spoke to the Manufacturing Chemists' Association, lauding the current system of regulation, and asserting with Sutherland that there was no need for the proposed Miller Amendment. Commenting on this speech, *Agricultural Chemicals* said that "unlike the Delaney hearings, wherein witnesses were subjected to cross-examination, the Association heard a lot of good talk straight from the shoulder."[42]

Economic entomologists had also hitched their wagon to the star of the new insecticides and were willing to defend them and the current regulations regarding their use. A good example of their rhetoric is Knipling's address to the American Association of economic Entomologists in 1953: "The Greater Hazard, Insects or Insecticides." He presented two alternatives: either continued use of insecticides under the direction of agricultural scientists, or a retreat to organic farming, starvation, and epidemic disease. He strongly defended the record of the newer chemicals, noting that DDT had saved 100,000,000 lives and that another of the newer insecticides, chlordane, which had been in use since 1946, had caused only one death. He urged his fellow entomologists to take care that chemicals were used carefully, but he saw no real danger. Chemicals were the wave of the future; "[t]he rapidity of future progress will depend in large measure on how well the research of entomologists, chemists, and toxicologists is coordinated."[43] The speech presented the standard defense of the new insecticides and showed the depth of commitment by entomologists. Use of insecticides, they came to say, was a subject on which they, and only they, were competent, and the stakes in terms of human life and food were too great to allow dabbling or overcautious regulation by interested but uninformed outsiders. Chemicals, too, became the only method of control. Funding for alternative means of checking pests declined in the wake of DDT and, despite scattered protests, entomology became, more and more, applied chemistry.[44]

Economic entomologists' increased dependence on chemicals and unswerving defense of them was not the only result of the introduction of DDT. At least as important was the larger part manufacturers took in the defense of what had been the farmers' position on regulation. During the 1930s, when lead and arsenic residues had been the problem, the opposition to government regulation and to the FDA's tolerance levels had come primarily from farmers and produce-marketing associations, such as the International Apple Association. Since the early 1950s, the main defenders of the current regulations have been the chemical manufacturers' lobbies—the Manufacturing Chemists' Association and, especially, the National Agricultural Chemicals Association. The rise of the NACA and its prominent role in the defense of DDT—its Industry Task Force for DDT was the chemical's principal defender in the late 1960s—reflects the increased concern of manufacturers over insecticide use in the period after World War II, the result of the enormous growth of the market. In 1939, for example, total pesticide sales (including insecticides, rodenticides, and fungicides) had been $40,000,000; in 1954 sales were $260,000,000, the result of the introduction of DDT, 2, 4-D, 1080, and a host of other compounds.[45] Agricultural chemicals were big business, and big business dominated by big companies. Dupont, Hercules, Olin Mathieson, and others began manufacturing DDT during the war, and after the war they continued to dominate the market. Fewer than ten companies supplied the vast bulk of DDT, and they took a strong interest in shaping government regulations. The trade journal *Agricultural Chemicals* began

publication in 1946 and at once began to serve as the NACA's
vehicle to unite the industry. The NACA was active in the
revision of the Insecticide Act of 1910 (the 1947 Federal
Insecticide, Fungicide, and Rodenticide Act) and in the drive
for uniform state regulation.[46]

While economic entomologists, insecticide manufacturers,
and government officials lobbied vigorously for their points
of view, the public took little interest in the matter. The
debate over the Miller Amendment, like the struggle over the
1938 Food, Drug, and Cosmetic Law, remained in closed
professional circles, a matter for the affected industries and
agencies. The most important reason was the general presumption, encouraged by DDT's apparently perfect safety
record, that the chemical was harmless. No one died from
DDT poisoning; no one even got sick. Long technical discussions of chronic toxicity, lesions on rat livers, and parts per
million in milk did nothing to disturb this presumption; they
only made the debate seem complex, abstract, and boring.
There were no simple, dramatic issues; there was no Harvey
Wiley to rouse the public. DDT also gained support from the
public perception, still widespread in the 1950s, that science
and technology were unmixed blessings. Strontium-90, pollution, and various chemical carcinogens were to make people
more skeptical, but not for another decade. Finally, no one
wanted too much publicity. Because of previous food-
poisoning scares and the public overreaction, the pesticide
industry believed that publicity would be more dangerous
than useful and the agencies, remembering their experience in
the prewar period, were reluctant to draw the criticism of the
lobbyists and their Congressional allies.

With the end of the Delaney hearings and the passage of the
Miller Amendment in 1954 the controversy over the new
pesticides seemed over, with the industry and the agricultural
community the winners. Although the FDA received new
powers of regulation, it did little to curb the use of the new
insecticides, and the major studies of the effects of DDT in
man were carried out by Wayland J. Hayes, Jr. within the

Public Health Service's framework. When insecticides did
return to the political arena, in 1962, it was as a result of the
powerful appeal of Rachel Carson's *Silent Spring*. By then
many people were becoming concerned, not with the effect of
DDT on public health, but with its impact on the environment, and scientists were becoming increasingly skeptical of
the harmlessness of the ubiquitous residues.

CHAPTER 4

Wildlife and DDT

MEDICAL SCIENTISTS could fit DDT into a familiar framework
—they had had experience dealing with insecticides—but
wildlife biologists could not. Earlier insecticides had been
confined by their high cost and low efficiency, to farms
and orchards. DDT's combination of high toxicity to many
insects, low mammalian toxicity, low cost, and suitability
for aerial spraying invited its use in areas that had been,
before World War II, free of insecticides. As soon as the
war was over, foresters and public health officials began
spraying millions of acres of swamps, forests, and suburbs.[1]
Wildlife biologists had to assess the effects of such sprays on
the environment, assess them before there was any permanent
damage (if DDT did turn out to be dangerous), and assess
them without any previous experience with such programs.
Like most scientific investigations, this one proceeded by fits
and starts, through false turns, unexpected discoveries, and
much hard work, and it went on without much public interest
and support. Not until the publication of *Silent Spring* in 1962
did the public begin to take an active interest in the problems
that the wide use of persistent pesticides might pose to
wildlife.[2]

Some of the earlier tests with DDT had been done in
forests—to assess the usefulness of the new insecticide against
the gypsy moth and the spruce budworm—and wildlife
biologists had been called to study the effects of the sprays.
Their role, though, was largely limited to assessing the
immediate mortality on susceptible organisms, setting down
rules to minimize the toll of birds and fish. By the time DDT
was available for general civilian use, there was a large body
of knowledge to aid economic entomologists in planning safe
campaigns. By February 1946 two government biologists,

Clarence Cottam and Elmer Higgins, could point to studies done in twelve states and Canada showing the dangers of DDT and the ways to minimize damage to the environment. Heavy or carelessly applied sprays, they warned, caused extensive mortality among fish and birds.[3] For example, two investigators had found that a single application of DDT to a Pennsylvania forest at the rate of five pounds per acre "resulted in the death of birds and a conspicuous reduction in the breeding population."[4] The use of DDT should therefore be restricted to the control of serious pests, they said. When it was necessary, it should be used with caution and under conditions that would minimize danger to wildlife. The smallest possible doses should always be used, sprays never applied to bodies of water, and strips left unsprayed as wildlife refuges.[5]

Evidence from field studies was not the only indication that DDT might have serious effects in the environment. Early tests had also shown that DDT taken into the body accumulated in the fat tissues and was passed on in milk.[6] Although no one believed that the chemical posed a danger simply because of these properties, many thought that there was a possibility that DDT might become a serious problem. Economic entomologists, as well as wildlife biologists, were cognizant of this danger. Shortly before Clay Lyle's call for the eradication of the housefly in 1947, the American Association of Economic Entomologists' Committee on the Relation of Entomology to Conservation took a somewhat more cautious view of the new chemicals. Noting the possibilities of chronic poisoning and the concentration of DDT in fat, the committee urged *"immediate experiments to test . . . the effects of various other organic chemicals that promise to become important insecticides,"* and urged that they *"be used cautiously until their actions on other forms of life and on the soil have been determined."*[7] Even George Decker, although convinced that DDT was not a hazard to humans, still thought it might be a hazard to other forms of life. In 1950 he warned that chemicals were not a cure-all and said that "many of the new insecticides can and often do upset the biological balance of an area and while promoting more effective control of one pest we produce an equally or even more destructive outbreak of some other lesser pest. . . ."[8] That same year Paul DeBach, a California entomologist working on biological control, pointed out just such a problem. Natural control of scale insects in California citrus groves had been disrupted by the use of DDT, which killed the natural enemies but not the scale.[9]

Like the debate over human health and DDT, the potential problems involving DDT and wildlife remained largely a matter for professional, not public, concern. Even the National Audubon Society was, until the mid-1950s, hardly bothered by the subject. President John T. Baker issued a statement in March 1946 urging caution in the use of DDT and warning that it might turn out to have long-lasting adverse effects—a warning repeated in 1949—but conservationists did not go beyond pro forma expressions of concern. Nor, until the mid-1950s, was there much investigation of the long-term effects of DDT and similar chemicals or even of their presence and distribution in the environment. Why? One reason was the novelty of the situation. Never before had a chemical been so widely and so heavily applied to such a large part of the earth. DDT was also unlike other insecticides, and until scientists became aware of its persistence, mobility, and bioconcentration through food chains, as well as the effects of residues on susceptible species, there seemed no need to engage in extensive environmental monitoring or to look for it far from the point of application. Another reason for the relative lack of interest in the long-range effects of pesticides was the presence of more pressing problems— habitat destruction, human disturbance, pollution, and the threatened extinction of species. Wildlife biologists had limited funds, and they concentrated on the most pressing problems they faced. Finally, the nature of DDT's effects in the environment delayed recognition of the problem. In small doses DDT residues affected hormonal balances; this effect, expressed in the reproductive failure of several species of birds of prey, was a key element in the environmental case against

DDT in the late 1960s. In large doses, though, DDT was a central nervous system poison, and this was the most obvious effect. Until scientists found evidence of reproductive failure connected with residues, they assumed that poisoning would be the major problem.

During the 1950s the most important element in attracting public and scientific attention to the hazards of the chlorinated hydrocarbon insecticides was the immediate mortality from large-scale or intense spraying programs. DDT seemed an ideal material for extensive spraying. The USDA planned several campaigns to eliminate or reduce imported pests, and it recommended DDT for these and for a wide variety of other, local, uses. The campaigns to eradicate the gypsy moth from the Northeast and the fire ant from the Southeast, and the use of DDT to control the insect vector of Dutch elm disease were the most important ones. The high toll of nontarget organisms and environmental damage stirred public interest in, and scientific investigation of, DDT far more effectively than any scientific warnings. The Dutch elm disease control program was the most visible of these object lessons, for the spraying was done in towns throughout the East and the Midwest; people could see the dead robins on their lawns, count the cost of the local program, and attempt to weigh inexpensive protection for their trees against the morning chorus of song. It was this program that gave Rachel Carson the image of the silent town in *Silent Spring*.

Dutch elm disease became established in this country near Cleveland, Ohio around 1930, and there was an independent outbreak in the East a few years later.[10] It spread rapidly through the elm–lined streets of eastern and midwestern towns threatening to wipe out the elm as thoroughly as chestnut blight had killed the American chestnut. There was no cure, and the only effective control measure was kill the bark beetles that transmitted the disease. Because elms stood over streets, homes, and playgrounds, control with arsenicals was impractical. Sanitation, the removal of dead and dying elms to reduce the beetle's breeding ground, was the only measure available before DDT, but it was expensive, and, in areas with large stands of low-value forest elms, economically impossible.[11]

There seemed a good chance that DDT might save the elms or at least slow the spread of the disease, giving scientists time to breed resistant strains or find a cure, and in 1947 the Bureau of Entomology and Plant Quarantine began an experimental spray program in Princeton, New Jersey. It treated elms in a selected area with two sprays a year, one a 2 percent, the other a 1 percent, DDT-water emulsion. The concentrated spray, four and a half pounds of DDT per tree per year, and the density of suburban elms meant that the level of DDT was far above that of any other control program. Almost as soon as the spraying began residents complained that birds were dying. The bureau at first dismissed these reports, but tests showed that there was a relationship between the sprays and bird mortality. Fifteen of twenty-six dead birds found in the area had DDT in their tissues, including ten immature specimens that could only have accumulated the residues there. A population survey was even more disturbing; nestling survival was only 44 percent in the sprayed area, compared with 71 percent in an adjoining unsprayed tract. The next year disclosed 122 dead or dying birds of twenty-seven species, and residents found eighty-one more specimens.[12]

Changing the sprays did not solve the problem; a survey of the Dutch elm disease control program on the University of Illinois campus at Urbana between 1949 and 1953 showed a pattern similar to that at Princeton, and other, more disturbing, problems. The 1,400 elms had been sprayed twice a year with a 6 percent DDT-water emulsion, at the rate of one to one and a half pounds of DDT per tree per spray. The first spray, in the fall of 1949, apparently caused little immediate mortality, but complaints about bird deaths began in the spring of 1950, before the second spraying. Analysis revealed concentrations of DDT and DDE in robins' brains ranging up to 252 and 139 ppm respectively, with a median lethal concentration of about 50 ppm DDT. Analysis of earthworms from the area showed that they were contaminated with DDT that had fallen on the soil or been in dead leaves, and this, together with the deaths before the spring sprays, suggested that indirect as well as direct poisoning was occurring.[13]

The BEPQ continued to recommend DDT, and communities continued to use it. Sanitation, according to the bureau's experts, was expensive, not as effective as chemicals, and in many cases impractical, either because of reservoirs of infection—stands of low-value wild elms—or because extensive intercommunity cooperation was necessary to ensure sanitation over a large enough area to be effective.[14] Methoxychlor, which became available in the early 1950s, was less toxic to birds, but it was four times as expensive as DDT.

As Dutch elm disease spread, states began to plan coordinated programs. The disease appeared in Michigan in 1950, and in 1953 the Plant Industry Division of the Michigan Department of Agriculture began a coordinated program of inspection and control in the infested areas using the USDA's recommended sprays. Despite this work, over 100 communities in the state reported Dutch elm disease in 1959.[15] The first report from Wisconsin came in 1956, and there were sixty-two confirmations by the end of the year. Some cities and towns began spraying that year and others quickly followed.

Protests generally began at the local level, as conservationists and birdwatchers saw the effect of the sprays and began to look for ways to stop the use of DDT. One of these protesters was Lorrie Otto of Milwaukee, who helped bring the Environmental Defense Fund to Wisconsin in 1968. Like many people, Otto had initially given little thought to potential problems with insecticides; she assumed that they were safe and only gradually changed her mind. In 1950, just after moving to Milwaukee, she hired a nurseryman to spray for spruce-bud gall. He used DDT, and sprayed everything until the milky emulsion "dripped from the trees." It was, he assured her, harmless and would get rid of the mosquitoes as well. She noticed that birds no longer frequented her backyard but decided that the smell or the sticky deposit left on the leaves had repelled them. The next spring she and her children found several robins dead or dying in tremors. She thought it was some bird disease, even after someone mentioned to her that it sounded like DDT poisoning. Gradually, other events broke down her scepticism about the harmlessness of DDT. Guiding Girl Scouts and garden club groups through the woods near her home, she found more tremoring birds, not only robins, but herons, woodcock and other species. Her children also noticed changes. Frogs and fish gradually disappeared from nearby streams, and the salamanders in the woods became scarce, then disappeared.[16]

Silent Spring had little to do with Otto's convictions; by the time it was published she had reached the same conclusions and found the book too depressing and too familiar to finish. By the mid-1960s she was ready to fight. In 1967, for instance, she brought a dead robin and a dead bat to hearings in Madison on banning DDT and laid them on the table. When Milwaukee and its suburbs, including her own town of Bayside, began to plan a program of DDT spraying for Dutch elm disease, she took immediate action. She knew Joseph Hickey, an ornithologist at the University of Wisconsin, but Hickey was no activist. An acquaintance put her in touch with Charles F. Wurster, Jr., a young biology professor at the State University of New York at Stony Brook who had just completed a paper on the effects of DDT sprays on birds. She wrote to Wurster and he sent her a copy of his article and a letter of encouragement, beginning a correspondence that lasted through the Madison hearing.[17]

She first went to a village meeting in Bayside, armed with a "lap-full of documents" Wurster had sent her and the conviction that people just did not understand the situation and would stop spraying when they learned the facts. The village president seemed sceptical, but he agreed to appoint a committee. It consisted of Lorrie Otto, the village clerk, one of the board members, and another man, not then on the board but later president. At the meeting the other members refused to read her articles, dismissed her fears as nonsense, and warned her not to make trouble. Her reception at other meetings and in other villages was much the same, and she had no success in stopping the sprays. Her problems were common ones among citizens trying to fight pesticide programs at the local level. Confronted with a housewife or a group of citizens, a pile of scientific articles, and a complaint about pesticides, most agencies and elected officials reacted with indignation, anger, or disinterest. Even when they did not dismiss complaints out of hand, officials were reluctant to enter into a discussion of the issues involved. Few were scientists, and even fewer were interested in the issue; they accepted the guidance of the USDA and the state.

The dead robins attracted not only "nature-lovers" but scientists. One was George J. Wallace, an ornithology professor at Michigan State University who became involved when the Michigan Department of Agriculture began spraying on the East Lansing campus. There was a small experimental program in 1954, an expanded one in 1955, and a full treatment of all campus elms twice a year in 1956.[18] The sprays decimated songbird populations on campus. By 1957, Wallace found, an estimated pre-spraying population of at least one pair of robins per acre had shrunk to "a few scattered adults and one young," and in 1958 and 1959 robins arriving on campus in the spring died or were gone by the end of June.[19] Other birds were also affected. By 1958, populations of 49 of 77 formerly abundant summer residents had been sharply reduced or eliminated.[20]

Wallace was seriously disturbed, and he undertook several studies to determine the extent and cause of the mortality. Supported largely by grants from private foundations, he conducted laboratory tests and field surveys. In the spring of 1960 he and his graduate students took a weekly census of the campus robins. They found that as birds arrived from the South the population rose to a maximum of eighteen pairs on the hundred-acre campus in mid-April, then declined rapidly. Only three pairs were left by the end of May. From letters, personal communications, and newspapers, Wallace also found and listed indications of bird mortality from around the state. Nearly all communities with spray programs reported severe declines in resident bird populations, particularly among insectivorous and ground-feeding species. His informants had found 94 species of dead birds, and chemical analysis of available specimens of 41 species showed that 34 of them had DDT residues in their tissues. There was a high correlation between high DDT levels and death. Dead and dying birds generally had lethal amounts in the brain, and there was contamination in eggs in deserted nests, embryos, and one dead nestling. Birds collected outside the area, on the other hand, had lower levels of DDT or none at all.[21] Although the agencies administering the program had modified the recommendations and spray formulations, they had not eliminated the high toll of nontarget organisms; the sprays were seriously affecting the breeding population of a large number of birds. Without judging the effectiveness of the program against bark beetles, Wallace was "inclined to question the whole program, as currently conducted, on ecological grounds. Any program which destroys 80 or more species of birds and unknown numbers of beneficial predatory and parasitic insects needs further study."[22]

His warning had little effect on either the public or the USDA. Even had he been willing and able to mount a large campaign against the use of DDT for Dutch elm disease control, it is unlikely that Wallace would have affected policy at that time. The USDA had never been particularly concerned about what it viewed as regrettable—but unavoidable and temporary—losses of wildlife. The public, faced with what seemed the certain loss of so many shade trees, was willing to accept the death of some birds, particularly since the losses did not seem serious. Robins still appeared, particularly in the early spring, and their failure to raise young went largely unnoticed.

Other scientists were also troubled. One was Charles Wurster, who went on to become a key figure in the Environmental Defense Fund, another was Joseph J. Hickey. Wurster played a key role in the presentation of the environmentalists' case against DDT; Hickey, though much less of an activist, played an important part in assembling the evidence.

Hickey did not suddenly come to a realization that DDT was environmentally damaging; his convictions grew slowly over a period of years and were the result of the accumulation of data and the testing of hypotheses. Even when he became convinced of the dangers of persistent pesticides, he was not an outspoken campaigner. He did not even regard himself as a conservationist. "I'm a scientist," he said, "not a conservationist."[23] The story of his gradual conversion to opposition to DDT is a good example of the growth of the scientific opposition to persistent pesticides and shows as well some of the important steps in the development of the environmentalists' case against DDT.

A scarcity of birds first made Hickey aware of the situation. In 1957 the Illinois Audubon Society charged that sprays were killing birds in that state, and the first sprays in Wisconsin caused a wave of public protests. Hickey was not seriously disturbed; he knew that most pesticides killed some birds and had no reason to suspect that DDT was different or that the mortality was serious. He dismissed the charges that the sprays were decimating bird populations; housewives were not scientists.[24] Then, in July 1958 he attended a funeral in a small Illinois town. There he noticed a mulberry tree. It was full of ripe fruit, but there was not a bird in sight. At the cemetery, outside town, he did see birds. He questioned the townspeople. Yes, they said, there had been birds in the spring, but they had gone; they must have traveled farther north. Yes, the town had been sprayed for Dutch elm disease.

Hickey decided to investigate. That fall he tried to determine the lethal dose for DDT in robins, and the next year he obtained a Department of the Interior grant to study the effects of the sprays on urban songbird populations. He selected six Wisconsin communities that were using DDT and five that were not. Periodic surveys during the breeding season showed that population density of robins was inversely proportional to the amount of spray and that there was a positive correlation between the sprays and the bird deaths. In Janesville, Wauwatosa, and Shorewood, communities that had been sprayed for three years, robin populations were respectively 31, 68, and 98 percent below that of comparable unsprayed communities. "Shorewood," Hickey said later, "had experienced *Silent Spring*."[25]

In 1959 he saw the effects of these sprays in Madison, when the elms on the University of Wisconsin campus were treated. Bird deaths began with the onset of spraying, peaked about three weeks later, and slowly declined. After a careful census of the sprayed area, Hickey estimated that there had been a minimum kill of 86 to 88 percent of the nesting birds. A second spraying in 1960 produced similar casualties and long-lasting results. The yellow warbler population only slowly returned to normal, and the screech owls that had nested in front of Birge Hall had disappeared.[26] Hickey was by now convinced that Dutch elm disease control sprays were a menace to resident and breeding populations, but he was not sure about other uses of DDT.

While Hickey was doing this, two other campaigns were rousing public and scientific interest in the side-effects of these chemicals: the use of DDT in an attempt to wipe out the gypsy moth, and the mass spreading of dieldrin to kill the fire ant. One was a renewal of an old effort to rid the forest of an important pest, the other a first attempt to do something about what had been a relatively minor pest. Work against the gypsy moth began first. After the failure of the biological control campaign just before World War I, entomologists had adopted a series of stopgap measures, designed to limit the moth's spread until they found some good means of controlling it. Since 1922 the major effort had been to hold a barrier zone, fixed along the Hudson River.[27] During the 1930s the USDA had conducted a series of experiments, using airplanes and autogiros, to see if forest spraying was practical. It would be, the agency concluded, if there were some insecticide better than lead arsenate. DDT was the obvious candidate. By 1946 USDA workers were beginning to talk about the promise of the new insecticide; by 1952 the agency had begun to list eradication as the ultimate goal; by 1956 the entomology research branch of the Agricultural Research Service (successor to the Bureau of Entomology and Plant Quarantine) and the Forest Service had wiped out over 3,700,000 acres of outlying infestation and were ready for a full eradication campaign. In April, the federal agencies secured state approval for a cooperative program to push the moth back into New England and, if that was successful, to try to wipe out the main body of the infestation.[28]

Plans for the first year called for spraying over 3,000,000 acres, most of it in New York. Much of the area, particularly on Long Island and in eastern New York, was farmland and suburb, and public complaints began almost as soon as the planes took off. Most were about nuisances—cars dotted with oil, swimming pools covered with scum, and yards and houses sprayed—but there were other, more serious problems. DDT on pastures and cattle contaminated milk, rendering it unfit for sale, and organic gardeners also faced serious losses. New York Commissioner of Conservation Sharon J. Mauhs reported that there had been significant fish kills in Orange, Ulster, and Sullivan Counties, which caused Governor Harriman to protest to the Secretary of Agriculture. The water commissioners of New York, fearing contamination of the city's water supply, attempted to ban all spraying within 500 feet of any reservoir.[29]

Public opposition went beyond protests; on May 8 several Long Island residents, led by Robert Cushman Murphy, curator-emeritus of the Museum of Natural History, filed suit in federal district court against Ezra Taft Benson, Secretary of Agriculture, Lloyd Butler (area supervisor of the Pest Control Branch of the Agricultural Research Service), and Daniel Carey (Agriculture Commissioner of the State of New York). They sought an injunction restraining the defendants from carrying out the DDT spray program, alleging that it might impair their health and endanger their lives. After a preliminary hearing, Judge Byers denied a writ but left the way open for a hearing on a permanent injunction.[30] The trial, which lasted from 11 February to 5 March 1958, showed the weakness of the case against DDT. Chief witnesses for the plaintiffs were Murphy and Dr. Malcolm Hargreaves, head of the hematology division of the Mayo Clinic. Neither had a

conclusive case. Murphy testified about the effects of DDT on
the marsh near his Long Island home. Apart from immediate
damage, he warned, such sprays could seriously affect the
balance of nature. He pointed out, as an example, that crabs
ate starfish eggs. Since starfish were predators on clams and
oysters, the death of crabs from the sprays could have serious
effects on the shellfish population, an important part of the
local economy. He did not, however, show that the sprays in
use did affect the crab population. Hargreaves's testimony
was more sensational, but just as theoretical. He contended
that DDT was linked to cases of various blood diseases—
including leukemia and aplastic anemia—in genetically susceptible individuals and that it posed a serious hazard to these
people. The evidence, though, was hardly conclusive. He
could produce neither a mechanism to account for DDT's
alleged action nor laboratory studies to support his clinical
cases. Much of his testimony was similar to that heard by the
Delaney Committee, evidence that indicated the need for
caution and further research, but which did not prove that
environmental levels of DDT were harmful.[31]

The defendants presented essentially the same case they had
given before the Delaney Committee and on which they were
to rely at other hearings. DDT, they said, was harmless to
man; government studies had shown it to be so. If there were
any long-term problems, they would be found by studies of
people exposed to high levels of the chemical and the government was carrying on this research. The star witness was
Wayland Hayes, who gave DDT a clean bill of health. He
cited his studies on convict volunteers who had ingested
massive doses of the chemical and the continuing research on
the health of workers in DDT-manufacturing plants. In both
groups, Hayes said, there were elevated levels of DDT in the
blood and the fat, but no pathology that could be traced to the
ingestion of DDT.[32]

The case did not attract much attention and ended quietly.
Judge Bruchhausen accepted Hayes's work on DDT and
dismissed Hargreaves's studies and warnings. The plaintiff's
major complaint, he said, was nuisance, and this was not a

sufficiently serious ground to stop a program that promised great benefits to the community.[33] The case—the first serious challenge to a spraying program on the grounds of public health and environmental safety—died without issue. It broke no new legal ground, asserted no public rights in the environment, and did not raise serious questions about DDT. There was, it must be said, little solid evidence to link pesticides to human health problems or to show that DDT use, as distinct from misuse, was dangerous to wildlife. In 1958 scientists were suspicious of DDT, but wildlife biologists were not willing, as many were a decade later, to assert that the chemical was a danger to the environment. The weight of medical evidence, and opinion, was still on the side of DDT's defenders.

Even before the Long Island case came to trial, the USDA launched another "blitz" campaign, this one against the South American fire ant, a pest that had established itself in the Southeast. The ant had entered this country about 1918 but had remained in a small area around Mobile, Alabama until 1930, when a smaller, reddish form, possibly a new introduction, appeared. The new race displaced the old and during the next twenty years occupied much of the lower South. Its mounds, which reached densities of twenty-five to thirty per acre and stood three to five feet high, interfered with farm machinery, and the ants were a nuisance to farm workers and, occasionally, a serious hazard to small children (as were wasps and bees). It did not seem to be a major hazard. A USDA study of the ant, conducted in 1949, concluded that it was a minor pest that would, because of the climate, remain confined to the lower South. In 1956, though, ostensibly in response to local complaints about the ants, the department decided that the fire ant was a menace to man and livestock and prepared an extermination campaign, which began in late 1957.[34]

Announcement of the program brought vigorous protests from conservation groups and individual scientists. The National Wildlife Federation and the National Audubon Society both called for a halt to the work. The Conservation Foundation and the New York Zoological Society joined them, urging more research on the problem before any action was taken and stressing the still unknown effects of dieldrin. Three Harvard biologists wrote a letter to the *New York Times* contending that the campaign would harm wildlife and might lead to serious ecological damage. The *Times* itself endorsed the Zoological Society's stand, and called for a reexamination of the department's current practices.[35] Critics charged that the new program combined the worst features of the gypsy moth campaign: indiscriminate spraying, the coverage of enormous areas, and little consideration of the side-effects with other elements of poor planning.

Questions could be, and were, raised about the need for ant control at all. Southerners, critics pointed out, had lived with the ant for forty years and as late as 1949 the USDA had paid no attention to it. Even the public demand for control, it seemed, might be manufactured. The key question, though, was why the USDA needed to undertake a blanket spraying campaign against the insect. It would be difficult to find an insect more susceptible to spot spraying; why spread pellets of a dieldrin-clay mixture over 20,000,000 to 30,000,000 acres? Dieldrin, a persistent chemical, posed obvious dangers to wildlife, either through direct poisoning or through the accumulation of residues in food chains. Although the USDA assured the public that the program had been planned to "do the least possible damage to wildlife and other insects," critics found this little comfort.[36]

The first sprays, in late 1957 and 1958, confirmed the critics' predictions. Fish, birds, livestock, and poultry died. Sportsmen noticed declines in gamebird populations, and Gulf fishermen found that crabs and shrimp had died from the runoff. There were long-range effects on wildlife; several species of snakes, lizards, and frogs were exterminated over parts of their ranges. John George, investigating the program for the Conservation Foundation, stated that the work had been undertaken without any real knowledge of the effects on wildlife, and George J. Wallace said that the fire ant eradication program was "one of the worst biological blunders that

man has ever made."[37] The fire ant program, like that against
the gypsy moth, quickly fell into disfavor; Alabama with-
drew matching funds in 1959, Florida in 1960, and other
southern states protested even when they agreed that action
had to be taken.

The campaigns stimulated public and scientific interest in
the effects of broadcast sprays on wildlife. On the national
level it triggered the first significant federal action on pesticides and wildlife; Congress, in 1958, appropriated funds for
a continuing study of pesticide residues in the environment,
the beginning of a national monitoring system that, by the
mid-1960s, helped to outline the dimensions of environmental contamination. The program, carried out under the direction of the Department of the Interior's Fish and Wildlife
Service, included grants to academic scientists to carry on
research, and studies such as Hickey's work on DDT in the
Great Lakes ecosystem in the early 1960s enjoyed its support.[38]

The eradication campaigns also alerted major conservation
organizations. The National Audubon Society, which had
hardly discussed the issue in the early 1950s, became increasingly concerned as the toll of wildlife mounted. John Baker,
Audubon's president, sent a letter to President Eisenhower,
asking for a halt to the gypsy moth control program. He cited
the unknown long-term effects of the chemical and problems
of estimating the true costs and the benefits and reiterated the
society's opposition to unrestricted aerial spraying. A month
later a press release described the large number of reports
about damage to fish and birds, warned about the indiscriminate application of poisonous chemicals from the air, and
asked for regulations shifting the burden of proof from the
public to the users of the chemicals. Meeting with the
Secretary of the Department of Health, Education, and
Welfare in December, Baker again stressed Audubon's opposition to the broadcast use of pesticides. He pointed to the
dangers of secondary poisoning and the research showing that
DDT caused reproductive failure in quail and pheasant.
Audubon opposed the fire ant program on the same grounds.

Testifying before the Subcommittee on Agriculture Appropriations in March, 1959, Baker said that there was no
justification for the program. "The threat to public health, let
alone people's continued ability to reproduce, is very great
and may well exceed that from radioactive fallout."[39]

The Audubon Society had come a long way from its
cautious warnings of 1946, but it was not ready either to
condemn pesticides, even persistent ones, or to engage in
direct confrontations with the USDA. Baker refused to
support the suit brought by Murphy's group on Long Island
against the gypsy moth program. His successor, Carl Buchheister, refused society aid to Rachel Carson's *Silent Spring*;
despite the recommendations of staff biologist Roland
Clement, the society decided not to publish the book. As late
as 1961, in fact, Buchheister listed the wilderness bill, the
national seashore, and preservation of wetlands ahead of a
new bill to coordinate pesticide use and registration as goals
for conservationists.[40]

Audubon's reluctance to forge ahead on this issue seems to
have been partly a result of its unwillingness to discard the
methods of lobbying and public education that had served it
well in the past and partly a reaction to the state of the
evidence. There was enough scientific information to show
that DDT would, in large doses, affect bird populations—
which brought Audubon's opposition to heavy sprays or
broadcast use of persistent materials—but not enough to
convince most observers that discriminate use, as distinct
from misuse, would be dangerous. Roland Clement remembers that a 1958 study showing DDT buildup in earthworms
convinced him that the chemical was dangerous, but he seems
to have been the only official in the Audubon Society who
went that far. There was opposition to Dutch elm sprays, and
even to the use of DDT, among the citizens organizing local
campaigns, but neither the Audubon administration nor the
scientists at national headquarters were willing, as yet, to
condemn DDT.[41]

The mortality from large spraying programs, though, was
not the only evidence that persistent pesticides might not be as

harmless as people had thought. During the late 1950s and early 1960s several studies suggested that there was an urgent need for more research on DDT and other such chemicals, that they might have unsuspected properties and could be causing environmental damage. Two points were particularly important—the discovery that food chains in the environment could concentrate DDT and the evidence that it might be interfering with avian reproduction. Bioconcentration had been known in a simple form from the early work on DDT, when studies had shown that cows fed DDT-sprayed forage concentrated the chemical in their fat and passed it on in the milk. What was disturbing was the discovery that more complex reactions might be taking place in the environment: plants accumulating residues from the soil or water, herbivorous fish further concentrating the chemical, and carnivorous fish and birds getting even higher levels with their food. The other discovery—that DDT could interfere with avian reproduction—was totally unsuspected, for DDT's acute effects were on the nervous system and the liver. But new evidence, suggestive but not conclusive, raised the possibility that DDT was a dangerous environmental contaminant, that even in concentrations too low to show pathological effects it could do harm.

Indications of reproductive problems came first, from work done at Patuxent Wildlife Research Center, the Fish and Wildlife Service's laboratory in Maryland. In 1951 scientists there reported that an experimental tract, sprayed with two pounds per acre of DDT a year for five years, had suffered a 26 percent decrease in the breeding population of birds.[42] Short-term studies of single sprays or censuses a year after a spray had not shown any diminution, but long-term observations suggested that repeated sprays might have entirely different effects. James Dewitt, a biologist at Patuxent, began feeding studies on quail, using DDT, dieldrin, aldrin, and endrin—all chlorinated hydrocarbon insecticides. The initial results were disturbing. He found that diets containing 0.002 percent DDT or 0.001 percent dieldrin affected both the hatching of eggs and the survival of the chicks that did hatch.[43] More research, published in 1958, confirmed these results. The data, he wrote, "support the conclusion that widespread, heavy, or improper use of cholorinated insecticides may damage bird populations by direct mortality or through impairment of reproductive functions."[44]

Field studies raised the possibility that DDT might be building up in food chains. In 1958 Barker published his work on the effects of DDT sprays on the University of Illinois campus. He had linked the death of songbirds to the use of DDT to control Dutch elm disease, but his studies went well beyond that. He had found DDT in campus earthworms, and it seemed likely that the sprays were being recycled back to the birds—from leaves to litter, to soil, to worm, and hence to robin. The timing of the mortality reinforced this conclusion. The first sprays had taken place in the fall of 1949, and the first bird deaths occurred in the early spring of 1950, before the second spraying.[45]

Two years later another study suggested that effects might not be limited to heavy use of DDT, that even low levels could become dangerous. This study, Hickey remembers, jolted him. The evidence came from an investigation into the death of western grebes at Clear Lake, California. Clear Lake, a 46,000-acre body of water 100 miles north of San Francisco, harbored a breeding colony of anout 1,000 pairs of western grebes. In the summer of 1949, though, the colony did not breed, and in December 1954 over 100 grebes died at the lake. More died the following spring and a third wave of mortality swept the colony in December, 1957. As a breeding population, it ceased to exist. Wildlife biologists, called in to investigate the deaths, were at a loss. None of the traditional causes seemed to fit. Almost as an afterthought they analyzed the visceral fat of two dead birds and found the "astounding level of 1,600 ppm" DDD.[46] DDD, a chemical closely related to DDT, had been used in the lake, but how could it have caused the problem? In September, 1948, the Lake County Mosquito Abatement District had applied DDD to the lake water in a dilution of one part in seventy million. (This gave a concentration of about fourteen parts per billion of DDD in the lake). The insecticides had been used to control the Clear Lake gnat, a small insect that was a nuisance to tourists, fishermen, and summer residents. The first application had killed about 99 percent of the larvae in the bottom muck and caused substantial reduction in the gnat population for the next few years. In September of 1954 and 1957 further applications, in an estimated concentration of twenty parts per billion, controlled the resurgent gnat population.[47] The use of the chemical correlated well with the mortality, but the levels seemed too low to harm anything, and indeed, there had been no immediate effects. The biologists concluded that contaminated fish were the source of the DDD and began sampling the lake biota. They found a pattern of concentration of insecticide residues in successive links of the food chain. A composite sample of plankton showed, for example, 5.3 ppm DDD, 265 times that of lake water. Herbivorous fish had twice the level of the plankton, and predaceous fish and birds had concentrations in their fat 85,000 times greater than that applied to the lake.[48]

These findings were, as Hickey pointed out, far more disturbing than the effects of the Dutch elm disease control program.[49] Deaths in the latter had occurred in small areas as the result of abnormally heavy sprays or through massive, direct contamination of the food supply. At Clear Lake, the pesticide had been applied in extremely low concentrations and had not caused immediate mortality. But, instead of degrading or otherwise disappearing, the pesticide residues had persisted, and their solubility in lipid tissue had caused a concentration up the food chain that had killed the grebes. Thus the question of the safety of persistent pesticides took an entirely new form. If these compounds survived and concentrated, did any application to the environment constitute a new addition to a permanent burden on all organisms? Entomologists and wildlife biologists alike had assumed that discriminate use of pesticides would cause no appreciable, permanent damage to wildlife or to stable ecological systems, but the Clear Lake results suggested that there might be no possible "discriminate" use.

The experience at Clear Lake was important in shaping research on insecticide residues in the environment, for it suggested new problems and new approaches. What were the chronic effects of exposure to insecticide residues? What levels now occurred in various ecosystems and how did the current levels affect organisms at the top of food chains? What populations, if any, were most seriously endangered? These questions lent new urgency to the environmental monitoring program; they also alerted biologists to the properties of DDT. Hickey was not the only one who began to consider DDT residues as a possible factor in environmental disturbances; many biologists now thought of insecticide residues as a subject to be investigated.

The reports of bioconcentration in the Clear Lake ecosystem did little to stir conservationists, but another issue did—the status of the bald eagle. In 1958 Charles Broley, a retired banker who had been banding and studying the bald eagles of Florida's west coast since 1940, made a startling suggestion. There had been, he said, an alarming increase in nesting failures among the eagles; about 80 percent of them were not raising young. Citing Dewitt's work at Patuxent, he suggested that the eagles had been made sterile by DDT residues in fish. Other observers, he noted, had found the same type of decline. Censuses of eagles along the upper Mississippi and at Hawk Mountain, Pennsylvania, showed an unusually small proportion of immature birds. Was DDT killing our national emblem?[50]

Had Broley simply been an amateur birdwatcher, scientists could have dismissed his work, but he was an accurate and careful observer. The Audubon Society, which published Broley's article, followed it with a commentary by Joseph C. Howell, a zoologist from the University of Nashville studying the eagles of Florida. Howell discounted pesticides. Although he admitted that the eagle population was falling, he blamed human disturbances and habitat destruction. He drew a sharp reply from Alexander Sprunt IV, an Audubon biologist, who chided Howell for neglecting the central issue, the possible sterility of the eagles induced by pesticide residues. Howell replied that he did not wish to comment on the subject because there was no data; Broley merely had a hypothesis.[51] The hypothesis, though, was alarming enough to stimulate further research. The Audubon Society enlisted its members in a drive to gather information about the eagle population, and it hired a biologist to make a survey of the area Broley had studied (Broley died in 1959). The Fish and Wildlife Service, too, joined the quest.[52]

These studies, coupled with rumors of other environmental problems, were beginning to have their effect on wildlife biologists. By about 1960, more of them were concerned that DDT and its metabolites might be serious environmental contaminants. There was, though, no agreement on either the seriousness of the problem or on what action to take. Some scientists, such as Roland Clement and George Wallace, were convinced that current DDT use was a danger to breeding populations and that its use should be restricted. Others, such as Hickey, were not willing to commit themselves. DDT, they thought, might be a problem, but much more work had to be done before that could established. On one thing the scientists could generally agree: their place was in the laboratory, not the public arena.[53]

One result of wildlife biologists' reluctance to involve themselves in public affairs, a stand that most scientists shared, was that pesticide policy remained in the hands of the groups that had for so long dominated the area: pesticide manufacturers, farmers, and the agricultural scientists in the Department of Agriculture and the state experiment stations. What debate there was on pesticide residues was carried on in a closed professional circle, with little effect on policy or on the public. Conservation organizations were, by 1960, interested but unable to rouse real public interest in the question. Not until 1962, with the publication of Rachel Carson's *Silent Spring*, did the situation change, and then it changed both drastically and permanently. *Silent Spring* marked a watershed, as the private, scientific debate became a public, political issue.

CHAPTER 5

Storm over *Silent Spring*

> It is not my contention that chemical insecticides must never be used. I do contend that we have put poisonous and biologically potent chemicals indiscriminately into the hands of persons largely or wholly ignorant of their potentials for harm. . . . The public must decide whether it wishes to continue on the present road, and it can do so only when in full possession of the facts.
>
> Rachel Carson

> The subject of pesticides should be put in the hands of professionals—it should be put back into the hands of the professionals in the College of Agriculture.
>
> Department of Entomology
> University of Wisconsin, Madison[1]

IN THE SUMMER of 1962 the *New Yorker* published three long excerpts from a forthcoming book by Rachel Carson, *Silent Spring*. By the time Houghton Mifflin published the full text in the fall, the book was the center of a noisy controversy that only increased as it became a best seller. It is not difficult to see why. For years everyone had assumed that pesticides were "safe"; Carson, in a well-written and apparently thoroughly documented book, said they were not. With the threat of nuclear war, she contended, the "central problem of our age" was the contamination of man and the environment with chemicals that disrupted ecosystems, killed animals and occasionally humans and that could "even penetrate the germ cells to shatter or alter the very material of heredity upon which the shape of the future depends."[2] Chemical insecticides, she charged, had been introduced into the environment and the human diet with "little or no advance investigation of their effect on soil, water, wildlife, and man himself. Future generations are unlikely to condone our lack of prudent concern for the integrity of the natural world that supports all life."[3]

Silent Spring was at once an exposé of the damage pesticides were doing and might do to man and the environment, a report on less harmful methods of insect control, and a plea for a changed attitude toward nature. It began with a "fable," an account of a "silent spring" in an American town where no birds sang, and no crickets chirped, a little world from which the voices of spring were absent. "No witchcraft, no enemy action had silenced the rebirth of life in this stricken world. The people had done it themselves."[4] They had done it with the miraculous chemical insecticides, spread liberally through the town. The rest of the book was devoted to showing how this silent spring might yet occur and what must be done to halt it. Carson discussed the chemistry and pharmacology of the new insecticides, their buildup in human tissues and their toxicity, and the evidence of their effects on water, soil, and vegetation. She passed on to the campaigns against the Japanese beetle, Dutch elm disease, the gypsy moth, and the fire ant, and then, in several chapters, told about the possible long-term effects on humans. The evidence, she said, indicated that DDT might be carcinogenic and mutagenic. Relying on the testimony and studies of Hargreaves and Hueper (who had appeared at the Delaney hearings), she presented both their evidence and their views on the need to protect the public from even low-level lifetime doses of the new chemicals. The crux of the argument was that it was "impossible to predict the effects of lifetime exposure to chemical and physical agents that are not part of the biological experience of man."[5]

Nature, though, was not helpless under this assault. Even now, she wrote, an avalanche was poised to descend on us. Insects were developing resistance, bringing to use newer and more deadly chemicals. The residues, distributed throughout the world, were causing unpredictable damage to the at once fragile and incredibly tough web of life that sustained us. Our salvation, she concluded, lay in a different approach to nature and different ways of controlling the pests that competed for our food. We must, she stressed, abandon the idea of conquering nature, of setting ourselves against the natural world and attempting to eliminate it where it interfered with our desires. We must instead seek to work with its processes. Ultimately we must realize we are part of it, citizens of the biological kingdom.

Despite the book's contributions to the environmental movement, *Silent Spring* was not primarily about the effect of chemicals in the environment. "It has always been my intention," Carson wrote in a progress report, "to give principal emphasis to the menace to human health, even though setting this within the general framework of disturbances of the basic ecology of all living things. . . . Evidence on this particular point outweighs by far . . . any other aspect of the problem."[6] Nor, despite the author's scientific training and the footnotes, was it based on original research. Carson drew on the scientific literature, the testimony of medical scientists from the Delaney hearings and from their later studies, work on the mortality among birds from large-scale spraying programs, and the increasing evidence suggesting serious environmental problems caused by DDT. From this admittedly fragmentary evidence, she suggested what might be the ultimate consequences of continued use of persistent pesticides and similar compounds. The result was a bold prediction, informed by a disciplined imagination and considerable literary skill. Although there is little doubt of the book's effect on the public, about Carson's talents as a writer—even critical reviewers acknowledged her graceful style—or the merits of the case, scientists disagreed about the value of the book. George Woodwell, an ecologist, believes that Carson used "the type of analysis and protection that we use in science all the time and often award Nobel Prizes for." Another ecologist, Joseph Hickey, thought the book was "full of truths, half-truths, and untruths."[7] Although agreeing with her basic point, Hickey was suspicious of the emotional content. As an analysis of the problems caused by pesticides in the environment, he said, he preferred Robert Rudd's *Pesticides and the Living Landscape*, published two years later.[8] The disagreements among scientists, and the lack of enthusiasm for *Silent Spring* among some who believed it was correct in its analysis, stems, however, more from a distaste for public controversy and emotional argument and from the belief that a scientist's place is in the laboratory than from any feeling (among wildlife biologists, at any rate) that Carson was misusing scientific studies. An emotional argument, though, was essential, for Carson did not write *Silent Spring* to tell scientists what was in the literature. She wrote it to rouse the public to what she considered a horrible danger.

Judged on these terms *Silent Spring* was a stunning success. It immediately made pesticide policy a matter for pubic debate and allowed the participation, in at least some fashion, of the public. Congress held hearings on government oversight of pesticides and President Kennedy asked his Science Advisory Committee to make a special study of the problem.[9] The contributions of the book, though, went well beyond the immediate issue of pesticides. Carson provided the first serious indictment of the indirect effects on the environment of a technology that seemed harmless. The effects of fallout, although shocking, were not completely unexpected; no one had held up the atomic bomb as a savior of humanity or even a nice thing. DDT, though, had seemed so harmless that it was routinely used around people, and children ran through the sprays from the trucks that fogged small towns and suburbs in the summer. If DDT did spread through the air, if it might cause cancer, if it killed millions of fish and birds, what was safe? What other things might not the generous and wonderful laboratories of the scientists harbor? In developing the case against DDT Carson also made what was the clearest case to that time of the central tenet of the environmental movement: that human action has become the dominant environmental influence on the rest of the planet, an influence that has spread to the farthest

reaches of the world. The spread of DDT residues provided
an object lesson, one driven home by the apparently limited
use of the chemical. Despite its confinement to settled countries, residues could be found in oceanic fish and even in
Antarctica. Finally, she provided an alternative ideology, a
replacement for the faith in science, technology, and power
that was the common wisdom. We must, she said, become
part of the natural world again, recognizing our ultimate
dependence on the processes of nature, and working with,
rather than against, them.

Carson's graceful prose did much to help her case, but
another factor added an appreciable, if unmeasurable,
weight—the slow sifting down from the sky of the radioactive products of atomic bomb tests from Nevada and from the
Pacific atolls. The image of contaminated rain, knowledge of
the slow circulation of strontium-90 from the stratosphere to
the grass to the cows to the milk American children drank
each day, the possibilities of leukemia and bone cancer—these
did more to make Americans suspicious of the utopian
dreams of technology and more to make them ready to
condemn environmental poisons than did anything else. The
public began to realize that the scientists, despite their awe-
some power, did not know the full consequences of their
action and could not predict the effects of the technological
marvels they had loosed on the world. The gradual discovery
of the nature of fallout contamination, the worldwide spread
of products from bomb tests conducted on isolated Pacific
atolls, and the fearful dangers to children set the stage for
Carson. By 1962 the public's naive faith in technology and in
technological solutions was beginning to erode; her audience
had come, if not to distrust technology, at least to be wary of
its possible dangers and disposed to consider charges of
danger from these miracles of modern living.

The enormous destructive power of the atomic bomb and
of the more powerful hydrogen bomb focused public and
scientific attention on the dangers of atomic war, but only in
the early 1950s did scientists and the public begin to consider
the dimensions of the threat to life and health posed by the

fallout from the weapons tests. The public first became aware of the possible dangers of fallout in the summer of 1954 when fallout from an American bomb test fell on the unsuspecting crew of the *Lucky Dragon*, a Japanese fishing vessel. By early fall, as the crew lay dying or recovering from radiation poisoning, and as worried Japanese turned away from fish, the American public became concerned. Concern turned to worry as research uncovered a more immediate problem; isotopes formed in the fireball were drifting to earth far from the point of explosion, contaminating the land with radioactivity.[10]

In the next three years extensive research and vigorous, occasionally sensational, reporting of the results gave the public a disturbing lesson about a small world. Even the vast reaches of the Pacific, it appeared, did not isolate us from the products of our ingenuity. The worst problem was strontium-90, a radioactive isotope formed in the fireball. The mushroom cloud, rising into the stratosphere, injected the material into the jet stream, which spread it throughout the world. Some of the isotopes were so short-lived that they decayed before they fell to earth, others were so long-lived that they posed little problem to man. Strontium-90, though, with a half-life of twenty-eight years, fell between the two groups. Worse, it began appearing in human food, particularly in milk. Strontium is chemically similar to calcium and cows grazing on contaminated pastures picked the chemical up and incorporated it into the milk they gave. Children drank the milk; the strontium settled in their bones, and the radiation (it was feared) might cause bone cancer or leukemia. Although only a small fraction of the radioactive strontium produced reached children, the chain still gave a new significance to the advertisements urging parents to make sure their children drank a quart of milk each day.[11]

By 1956, fallout from weapons tests had become a political issue; both presidential candidates had to deal with it, though neither sought to make capital of it. More important, fallout led to the formation of a movement to halt atmospheric testing, a movement that served, in some ways, as a prelude to the environmental movement. It helped to educate the public, to convince people that far-off actions could have consequences on their lives, that the earth was a closed system. The common attitude of "dump it and forget it" did not seem to work; Americans were forced to realize that the diluting effect of the atmosphere might not be enough. The test-ban movement also attracted some scientists and laymen who were later active in the environmental movement, including Barry Commoner, a biologist from St. Louis. Convinced that the public needed education on the issue of fallout and that scientists should participate in that education, Commoner was active in the Committee for Nuclear Information, and has since become one of the most active and prominent advocates of environmental action. The public discussion about fallout also introduced the public to other issues and problems that were an important part of the environmental movement—the difficulty of predicting the effects of new technologies, of estimating the hazard from low doses of hazardous substances, and the problems of carcinogenesis, mutagenesis, and teratogenesis. It even raised a now-familiar problem of government regulation. What agency should carry on the regulation and how did regulation conflict with promotion? Was it not a conflict of interest, one Congressman asked, for the Atomic Energy Commission to be developing atomic bombs and power plants and also monitoring the dangers of radiation?[12]

Silent Spring appeared when the test-ban issue was reaching a head, and, unfortunately for the chemical companies, just as another incident called into question the benefits of modern science and the adequacy of government regulation. In July 1962, a month after the *New Yorker* began publishing parts of *Silent Spring*, the *Washington Post* published a front-page article on a "modern American heroine," Dr. Frances Kelsey, whose "scepticism and stubbornness . . . prevented what could have been an appalling American tragedy, the birth of hundreds or indeed thousands of armless or legless children." Kelsey was the FDA physician whose insistence on more data on the safety of thalidomide prevented the general sale of that compound in the United States. Although the application was still pending—and the drug company was complaining about Kelsey—physicians established a link between thalidomide and birth deformities. Introduced in 1957 by Chemie Grunenthal of Stolberg-am-Rhein as part of a flu medicine, the compound proved to have unexpected side-effects. By late 1961 physicians in Germany and Australia independently traced an alarming increase in peculiar birth defects to the ingestion of the compound during the first trimester of pregnancy. It produced a wide range of distinctive deformities. Babies were born with no arms, their hands attached directly to the shoulders; some had no legs, their feet seemed to sprout from their hips; others had no limbs at all. There were, in addition, children with brain damage, deafness, blindness, epilepsy, and other disorders or deformities. There were, worldwide, some 8,000 deformed children in forty-six countries. The chemical was never sold in the United States, but people who received it in clinical trials or had taken it on a trip abroad bore deformed babies or were faced with the choice of abortion or the possibility of giving birth to a deformed child.[13]

Neither pesticide manufacturers nor users could, under the circumstances, assume that the public would ignore Carson's case. Three separate groups of critics quickly leaped to the defense of agricultural chemicals in general, and DDT in particular. The chemical industry and "agri-business" were concerned about the danger that a campaign against DDT would pose to other chemicals; they saw the problem in terms of public relations and the answer as improved public education to make everyone aware of DDT's benefits. Economic entomologists, though they appreciated the need for chemicals and the need for education, were more concerned about the reputation of their discipline. Many of them saw *Silent Spring* as an unprincipled attack on their professional standards. A third group rallied to the defense of DDT on principle. Professionally and financially independent of the controversy, these people saw *Silent Spring* as a reactionary book, as an attempt to undermine science and technology.

Carson, in their view, was a simplistic nature worshiper intent on subverting the progress of science that lay at the heart of Western civilization.

Each group had its own reasons for being concerned, but their defense can, without significant distortion, be treated as a single argument. Pesticides, they all stressed, were essential for modern agriculture and public health; we could not do without them. DDT (Carson's most conspicuous target) was, they said, safe for man. The experience of a quarter century, during which no one had died from DDT poisoning, and the research done by the Public Health Service, showed that there was no danger to the population at large. As for wildlife, there was certainly some mortality from heavy or careless spraying, but, properly applied by trained and careful personnel, DDT caused no lasting harm to animals, fish, or birds. Finally, *Silent Spring* was a distorted piece of work, a hysterical argument against science and technology. It would, one reviewer thought, appeal to "organic gardeners, the antifloride leaguers, the worshipers of 'natural foods' . . . and other pseudo-scientists."[14]

Pesticide manufacturers did not directly meet Carson's charges of damage to wildlife, nor did they attempt a serious analysis of the medical evidence of DDT's carcinogenicity (probably a wise move, given the ambiguity of the available data). They were primarily concerned with DDT as a public-relations problem and attempted to calm the public by presenting a simple, easily understandable case, backed by all the scientific authority they could muster. In 1962 the industry had an impressive case and an impressive number of scientists on its side. DDT poisoning was so rare as to be nonexistent; the chemical's acute toxicity was extremely low. DDT has been used in many countries, on millions of people, without noticable harm. Workers in DDT-manufacturing plants and convicts had endured high exposure without becoming sick, and there was no noticeable diminution of wildlife due to DDT use, except in areas that could be dismissed as unusual. Agricultural Chemicals had no trouble in assembling a group of scientists to answer Carson, and many, such as J. Gordon Edwards and Robert White-Stevens, ardently defended DDT before the public.[15]

The reasons for the industry's sharp reaction to *Silent Spring* can best be understood by considering an earlier incident, the cranberry scare of 1959. In 1957 the Department of Agriculture approved a new chemical, aminotriazole, for weed control in cranberry bogs, specifying that it be applied only after harvest. Farmers apparently did not follow these directions, for the FDA found and confiscated several lots of contaminated cranberries that year, storing them under refrigeration pending tests of the chemical's toxicity. Life-time studies on rats, completed in 1959, showed that aminotriazole caused cancer in the rats' thyroids. The 1958 Food Additives Amendment to the 1938 Food, Drug, and Cosmetic Act, usually called the Delaney Amendment, required the FDA to set a zero tolerance level for any chemical that caused cancer in any experimental animal.[16] The agency therefore destroyed the contaminated fruit and warned the growers of the situation. On 9 November 1959 Secretary of Health, Education, and Welfare Arthur S. Flemming held a press conference to announce the decision. He told reporters what had happened, including the information that aminotriazole caused "cancerous growths in the thyroid of rats." The FDA, he said, would check the entire cranberry crop, and, though no evidence of contamination had appeared that year, the public should not buy cranberries for Thanksgiving until the FDA had finished its work.[17]

Flemming had consulted the major growers before holding the press conference and had promised speedy action in clearing the crop before the holidays, but that did not dampen their reaction. Ambrose Stevens, executive vice-president of Ocean Spray Cranberries, Inc., decried the action. "We are shocked," he said, "that the United States Government has made public what we consider an inflammatory statement concerning possible contamination of cranberries by a weed-killer approved by the Department of Agriculture."[18] American Cyanamid and Amchem, the manufacturers of aminotriazole, issued statements denying that there was enough

weed-killer on the berries to cause cancer (something that
Flemming had never claimed).[19] The next day a meeting of
cranberry growers called for Flemming's resignation, saying
that his statement had been "irresponsible, ill-informed, and
ill-advised," Stevens issued another statement pointing out
that there was no evidence that aminotriazole caused cancer in
humans, and American Cyanamid criticized the FDA's toxicity studies.[20]

The public's response was strong and unequivocal. Sales of
cranberries, which should have been rising in anticipation of
the holiday, fell sharply. Grand Union, A&P, and Bohack
Supermarkets, among others, announced that they were
halting sales of cranberries, and some restaurants removed
them from their menus. In an attempt to reassure the public
George Olsson, an Ocean Spray executive, consumed a
handful of cranberries for the press cameras. The administration also sought to calm the public. Ezra Taft Benson,
Secretary of Agriculture, announced that his family would
have cranberries with their Thanksgiving dinner, and Vice-
President Richard M. Nixon ate four helpings of cranberries
at a dinner in Wisconsin Rapids, Wisconsin.[21]

The panic, for such it was, soon subsided. The FDA, by
using the cranberry growers' labs for testing and assigning
extra field agents to the job, managed to clear the crop before
the holiday season. There was no public reaction, no call for
investigation, and no new legislation; the incident passed
from the public mind. This was luck. All the ingredients of a
disaster had been present: incomplete tests of a chemical,
careless use, and nationwide distribution. The situation might
have been far worse. The pesticide industry, determined to
prevent a recurrence of this type of scare, began to devote
more time and attention to public relations, and when *Silent
Spring* appeared it swung into action.

The defense of DDT, though, was more than the defense of
an industry and its profits. Although some of the spokesmen
for the chemical doubtless had financial interests, most did
not, and to view the campaign for DDT in terms of eco-

nomics leaves unexplained the passions that the participants brought to the fight.[22] Economic entomologists were particularly incensed. They had devoted their professional lives to ending disease and want and felt that an attack on DDT impugned their professional values, standards, and actions.[23]

The closing of professional ranks and the entomologists' conception of their responsibility can be seen in the controversy over *Silent Spring* at the University of Wisconsin, Madison. During the first two weeks of October 1962, the campus radio station, WHA, read *Silent Spring* on the program "Chapter a Day," and Karl Schmidt, the announcer, devoted an entire program to reviews of the book, including the full text of a critical review by Ira L. Baldwin, a University of Wisconsin professor of bacteriology.[24] Two days later WHA presented a panel discussion, "Is the Use of Chemical Insecticides Harmful to the General Welfare?" Robert J. Dicke, chairman of the entomology department, was a member of the panel and had approved the selection of the members; Ira Baldwin was the moderator. It included another entomologist, an ornithologist (Joseph Hickey), a zoologist, and one specialist in oncology and cancer research. The members were, with the exception of Hickey, critical of *Silent Spring,* and they presented a strong defense of pesticides.[25]

In November the station presented a second discussion, "Insecticides and People," chaired by Karl Schmidt. Panel members were Grant Cottam and Hugh Iltis of the botany department, Van R. Potter, professor of oncology, Arthur D. Hasler of zoology, James F. Crow, medical genetics, and Aaron Ihde, who held a joint appointment in chemistry and history of science. These men were far more critical of both the pesticide industry and the applicators than the previous panel was. They did not condemn pesticides but agreed that there was a disturbing ignorance about the effects of the newer materials on man and other forms of life, and, among entomologists and manufacturers, a general lack of concern over effects on nontarget organisms. Scientists, they said, should neither ignore the problem nor limit public debate, and they rejected the contention that discussion would merely, and unnecessarily, alarm the public.[26]

Two months later a reporter for the (Madison) *Capital-Times* wrote a story, "The Uproar over Silent Spring," about the second discussion. It summarized the participants' views and criticized their opponents, noting that many of them were associated with chemical companies. The reporter quoted Ellsworth H. Fisher, an extension entomologist, but also gave a rebuttal and a strong defense of *Silent Spring*. In early March the entomology department replied with a nineteen-page "Critique of 'Insecticides and People,'" which had been "prepared by the University of Wisconsin Department of Entomology." (The manifesto, written by R. Keith Chapman, bore the names of all the members of the department; Chapman was not singled out as the author.[27] Privately circulated to the panelists, WHA, and various university administrators, it was, it declared, a response to the "unfortunate and undesirable situation" that had been brought to a head by the newspaper story, a situation "that has tended to degrade professors in the College of Agriculture, particularly those in the field of pesticides, has questioned the integrity, qualifications and intelligence of those workers and has interfered with their assigned educational programs." Karl Schmidt had, the critique said, "initiated the situation" by reading *Silent Spring,* and by his "biased and unethical handling of the ensuing controversy. He made every effort to surround the book with an aura of scientific authenticity and ignored the expressed opinions of persons on the campus who are best qualified to judge the scientific quality of such a book."[28]

The second panel discussion, it went on, had aggravated the problems. None of the participants was qualified to "speak to the public on the widescale uses and dangers of insecticides and other pesticides." This was shown by the "general lack of facts presented in their hour-long discussion, by the many false and misleading statements made, and by the fact that to the best of our knowledge not one of the panel members has ever applied a pesticide on a large scale and studied the effects of such an application." Further, the panelists had "inferred that we are irresponsible, unintelligent, lacking in professionalism, incapable of carrying out our assigned duties, and guilty of censorship in the area of dangerous knowledge."[29] The body of the paper dealt with some of the panelists' statements. It accused Schmidt and WHA of violating the rules laid down in their policy statement and the panelists of helping compound this damage to the university. Their "misinformation . . . has . . . degraded the public image of concerned responsible scientists in the College of Agriculture and has lessened their effectiveness. . . ." This could only encourage public alarm, frauds and cranks, and the "irresponsible (and degrading to the University) journalists such as the one" who had written the article.

The subject of pesticides, it said, "should be put back into the hands of the professionals in the College of Agriculture. . . . We agricultural scientists have been given the responsibility for making pest control recommendations."[30]

> The type of responsibility which pesticide workers must assume is something that some scientists in other areas on the campus apparently do not realize, do not appreciate, and possibly cannot even comprehend—having never been charged [with] such responsibilities themselves. As a result, these people possibly don't realize the enormous amount of damage and confusion that can result from their uninformed intrusion into the pesticide field or realize the dangers of making broad sweeping statements or loosely philosophizing about a subject in which their background information is so inadequate.[31]

Only those who had applied pesticides, it said, could judge them.

The critique is unusual only in so baldly stating the economic entomologists' position; in most cases the very emotional defenses of DDT were made by those not professionally concerned with the problem. That the entire department signed the critique—even though it did not allow public distribution of the document—indicates an unusual degree of interest and commitment.

The reaction was largely due to the professional ethos of the economic entomologists, built up over a long period of time. Since the emergence of economic entomology as a distinct applied science in the late nineteenth century, its practitioners had seen themselves as dedicated public servants, committed to controlling insect pests, rendering aid "to the distressed husbandman," and making every dollar of their appropriations yield some profit for him.[32] As early as the 1920s, when economic entomologists had decided that "the burden of control rests upon the use of insecticides," they had tied their professional hopes to chemicals and they had seen the new insecticide, DDT, as the fulfillment of those hopes. Reputation, career, and self-esteem were tied to the program of research that Carson had so clearly and vigorously condemned. If economic entomologists felt professionally and personally threatened by this argument, if they responded emotionally to the issue, there seems little reason to condemn them. They saw Carson's argument for what it was, an attack on the kind of insect control on which they had built a valued and valuable profession. If they mistook the attack on a method for an attack on the profession itself, the error is understandable.

The third group of critics, many of them people not professionally concerned with DDT, were less interested in the specific case Carson made than in its implications. The control of nature, she had written, "is a phrase conceived in arrogance, born of the Neanderthal age of biology and philosophy, when it was supposed that nature exists for the convenience of man."[33] She called for an end to the ideas that man can or should control nature and that man was not part of the natural world, ultimately limited by the need to retain the biological systems that, in Carson's view, supported all life. Acceptance of Carson's ideas, one critic wrote, would mean "the end of all human progress, reversion to a passive social state devoid of technology, scientific medicine, agriculture, sanitation. It means disease, epidemics, starvation, misery, and suffering."[34] The defense of DDT, then, became the defense of civilization.

It is in this call for a new attitude toward nature, for a new recognition of the possibly destructive effects of man's actions on the environment, that *Silent Spring* made its greatest contribution to the environmental movement. In DDT Carson identified a target that would serve as a focus for environmental concern, and her advocacy of a philosophy that saw man as part of the natural world helped direct the anger she roused. The campaign against DDT might well have been waged on grounds of self-interest (particularly given Carson's emphasis on human health); it was instead animated by a commitment to change people's ideas, not simply to undo the damage that had been done and to protect man, but to protect the environment around him.[35]

Man's relationship to nature, though, was not the main issue in the debate over *Silent Spring;* the question was how to regulate DDT. Sound policy could only be made on the basis of knowledge, and few were willing to accept without further ado the recommendations of one former government biologist, however well versed in the literature. The government began to call on the scientists. The President's Science Advisory Committee, headed by Jerome Weisner, prepared a report on the use of pesticides. The National Academy of Sciences—National Research Council wrote a three-part study of pest control and wildlife, and a subcommittee of the Senate Committee on Government Operations began a fifteen-month set of hearings on the environmental hazards posed by pesticides, the adequacy of federal oversight, and the extent of interagency cooperation.[36] The reactions and reports from these groups varied but there was a common core of caution.

Use of Pesticides, the report of the President's Committee, took a middle ground. The scientists did not agree with Carson's contention that DDT was a potential hazard to human health. There was, they thought, no evidence to prove that it was harmful to man. Nor did they entirely share her enthusiasm for nonchemical methods of control; chemicals were still essential for agriculture and public health. On the other hand, they thought that there were environmental problems connected with the wide use of these compounds and deficiencies in federal regulation. Although pesticides were applied to only limited areas, residues were found in ecosystems throughout the world. There was also considerable mortality among wildlife from some sprays; the report cited the toll among songbirds from the Dutch elm disease control sprays, the virtual extermination of young salmon in the Miramichi River in New Brunswick in 1954 and 1956, and the deaths of the grebes at Clear Lake.[37] There were also serious gaps in our knowledge of the effects of pesticides on man and wildlife and too little research on this subject. For example, though many pesticides were known to be synergistic (their effects in combination were not the sum of their separate effects) there had been little investigation of this phenomenon and its consequences to man. Federal regulation was also lax. The panel found that, although the USDA required stringent tests of a pesticide's efficacy, it did not require similar proof of its safety. There was no evidence on the "no effect" level of several important, persistent pesticides and inconclusive evidence on the chronic effects of pesticide exposure over a lifetime.[38]

Although the Weisner Report did not recommend the ultimate elimination of chemicals, even as a long-range goal, it did outline ways to reduce or eliminate the hazards associated with their use, ways that would bring major changes in pesticide policy. It recommended that programs be funded to assess the extent of present hazards and to guard against new ones through greater understanding of the behavior of pesticides in the environment. This would require a comprehensive data-gathering program on residue levels in both the general population and in occupationally exposed persons, an expansion of state pesticide-monitoring agencies, and a network of stations to check residue levels throughout the country in wildlife, fish, water, the soil, and man. The FDA, it said, ought to include more compounds in its studies of residues in the diet and to review their permissible residue levels. It also suggested a restructuring of government regulation of pesticides—clearer division of authority and greater cooperation among agencies and a transfer of responsibility for pesticide registration from the Department of Agriculture to the Department of Health, Education, and Welfare. Finally, it said, the "accretion of residues in the environment should be controlled by orderly reduction in the use of persistent pesticides. . . . Elimination of persistent toxic pesticides should be the goal.[39]

Although it did not endorse many of Carson's charges, particularly with regard to human health, the Weisner Report was controversial, for it envisioned a drastic shift in government regulation and control of pesticide use, with a corresponding change in the control over policy. To shift responsibility for pesticide registrations to the FDA, which then only controlled the level of residues on products reaching the market, would seriously upset agricultural and industrial power to influence regulation. It would also remove Congressional power over pesticide policy from the agricultural appropriations subcommittees. The extensive environmental monitoring of residues would give more authority to the Department of the Interior, whose Fish and Wildlife Service was actively engaged in studying pesticides in the environment, and it would expose current pesticide practices to the criticism of an experienced and sceptical group of scientists who already had serious misgivings about some current uses of the chemicals.

The report did more than add new agencies to the regulatory structure, it introduced new criteria for safety. It called for a continuing investigation of residues, which would give much more weight to long-term effects on people, including chronic problems, and to long-range effects in the environment, as distinct from immediate mortality. The elimination of persistent pesticides, though, was the most revolutionary goal. It would mean the end of a line of research pursued since World War II, when economic entomologists had put their faith in the new chemicals, almost to the exclusion of other methods of control.[40] As insects developed resistance to one chemical, another, entomologists assumed, would take its place. The Weisner Report called for an end to that idea, for a greater consideration of the side-effects of insect control in the environment.

The report of the National Research Council's Committee on Pest Control and Wildlife Relationships, on the other hand, did not recommend any drastic measures. The three-part study, made to "evaluate the problem objectively and to make proposals toward increasing the benefits of chemical pesticide use while minimizing the danger to wildlife,"[41] concentrated largely on the benefits of pesticide use, the necessity of insect control, and the dangers of misuse. Far from being an incisive scientific statement of the problem and of projected solutions, the report merely repeated old plati-tudes. Part of the problem may have been the makeup of the committee. Ira Baldwin, George Decker, Mitchell Zavon, Donald Spencer (of the NACA) served, as did Clarence Cottam, Ira Gabrielson (a noted conservationist), and others from various fields and of various views. Roland Clement, viewing the third section, on research needs, said that it lent "credence to the old saw that the committee method is the devil's preferred device for avoiding decision and preventing action." Clement's verdict, although that of a partisan, was not far from the mark. The report did not represent anything but the lowest common denominator, and their recom-mendations were obvious. Cottam, reportedly, was so furi-ous at the bland report that he held up publication of the third section for over a year.[42]

The Weisner Report appeared in time to become part of the controversy before the Ribicoff Committee, the subcommit-tee of the Committee on Government Operations that held hearings on pesticide use and interagency coordination on pesticides.[43] It was ostensibly interested in the efficiency of the federal government in dealing with information on pes-ticides, but there was no important legislation before the subcommittee, no significant scandal, no demand for a change in government—only the case so eloquently made by the former government biologist. As at other hearings, the

Senators heard from a variety of groups—industry and lobbies for industry, farmers, federal and state agents, and
private citizens representing various views. The views,
though usually not consciously coordinated, fell into a few
categories. There was a rough division between those who
believed that no changes needed to be made and those who felt
that federal regulation of pesticides had to be modified in light
of what was now known about the properties of the new
compounds. There was a similar division regarding human
health; one side said DDT was safe, the other said it was not.
As they had at the Delaney hearings, manufacturers and the
Department of Agriculture stood together, opposed by a
more heterogeneous group seeking changes.

The committee, of course, heard from Rachel Carson. The
evidence accumulated in the last year, she said, lent further
support to her case, and the Weisner Report had made many
recommendations she favored. The most important problem
was that of pesticide mobility. Much of the spray dumped
from airplanes never reached the ground but drifted with the
wind far from the point of application. Even that which did
settle on the crops or forest did not stay in the area. The
chemicals evaporated, codistilled with water, clung to parti-
cles of dust spread by the wind, or were concentrated by
living organisms that then spread through the world. Areas
within the Arctic Circle, far from any possible place that
used DDT, had measurable residues and even ocean fish had
accumulated the chemical. Faced with global contamination,
Carson said, we had to take action. She recommended, as a
minimum, limiting the use of insecticides and the number of
compounds approved. Citing the Weisner Report on physicians' general ignorance of the problem, she called for new
programs of medical research and education, and for a change
in pesticide policy, especially for the participation of other
agencies.[44]

Witnesses for the Department of Agriculture and the
pesticide manufacturers strongly disagreed with both Carson
and the Weisner report. Secretary of Agriculture Orville
Freeman defended the eradication programs, which Carson

had attacked and which the President's Committee had called unrealistic. They were, he said, useful, and the department planned to continue the work. There was adequate government regulation of pesticides, he went on, and excellent coordination with other agencies on pesticide studies. At the same time, Freeman stressed the USDA's willingness to change. The department supported the elimination of "registration under protest," a system that allowed a manufacturer to sell a product or products that had been rejected by the USDA until he had exhausted his legal appeals, unless there was a clear health hazard. More research, he agreed, was necessary on the long-range effects of residues in man and wildlife. As for persistent pesticides, the USDA was earnestly seeking to replace them; two-thirds of its research effort now went into biological controls, more specific chemicals, basic studies of insect pests, and new methods such as sex attractants and sterilization techniques.[45]

The strongest defenders of the status quo were representatives of the pesticide manufacturers, who asserted that there were no problems with current uses and no need for further legislation. Parke W. Brinkley, head of the National Agricultural Chemicals Association, repeated his charge that pesticides were already "one of the most regulated industries" in the country. "In our opinion," he said, "we do not need more controls as much as we need more basic scientific knowledge." He disagreed strongly with the President's Committee on the need to phase out persistent pesticides; they were necessary for public health. Brinkley feared that public hysteria would result in unnecessary government regulation, causing less effective controls of insect pests. "Carried to an emotional extreme, it would eliminate many effective and desirable products, free choice and free competition from the market place, and, perhaps, create by government dictate a virtual monopoly for one product."[46] George Lynn, head of the NACA's regulatory advisory committee, echoed Brinkley's fears and asked that authority over pesticides "reside in one person, and that person [be] . . . the Secretary of Agriculture. . . ."[47]

Besides the problem of who should regulate pesticides, the Senators had to consider the issue of public health and DDT, and on this issue they heard the same arguments voiced before the Delaney Committee thirteen years before—in some cases by the same people. The Weisner Committee had found that government research had not produced enough data, particularly that "necessary to predict the long-term, low-level toxicological effects for these chemicals on human beings."[48] Dr. Wilhelm Hueper, a specialist in environmental cancer at the National Cancer Institute, agreed. He pointed to indications that DDT was at least a potential human carcinogen, for tests on rats and dogs had produced cancer of the liver. Although there was not a one-to-one correlation between man and animal, there was a connection. What was needed, he said, were lifetime studies on experimental animals, studies that would include the effects during pregnancy and the health of the newborn. Dr. Malcolm Hargreaves of the Mayo Clinic agreed about the possible problems of DDT. Indeed, he was even more emphatic. After thirty years spent studying the problems of environmental medicine, he had concluded that most blood dyscrasias, including leukemias and lymphomas, were produced in genetically susceptible persons by exposure to environmental agents, including hydrocarbons. Like Rachel Carson, he believed that the great increase in the use of chemicals in the environment posed potentially serious health problems.[49]

Several physicians disputed these judgments. Dr. Charles Hine, who had appeared before the Delaney Committee in 1950, testified that there were no detectable ill effects from exposure to DDT residues, and he devoted some time to countering Hargreaves's evidence. The connection between pesticides and blood pathology, he said, was quite indirect and ill defined. There was no reason to suspect that DDT was the cause of these conditions. Dr. Mitchell Zavon, a Cincinnati public health officer, agreed. The best way to uncover the effects of pesticides, he said, was not by observing isolated cases, but by studying groups exposed to high concentrations of the chemicals for a long period of time. The U.S. Public

Health Service, he pointed out, was conducting such research and had found no problems traceable to DDT.[50]

The most effective advocate of this position was the man who was conducting the Public Health Service's studies, Dr. Wayland J. Hayes. Hayes had presented his views at the Delaney Committee hearings, and he repeated his case here with additional evidence. The Public Health Service, he said, had conducted studies to determine the effects of prolonged, high exposure to DDT. It had studied workers in a DDT-manufacturing plant and convict volunteers who had ingested daily doses of DDT for up to a year. Despite the high exposure, and blood and fat levels far above those in the general population, he said, there had been no detectable clinical effects traceable to the chemical. Hayes also went over the evidence for DDT's carcinogenicity and concluded that it did not implicate the insecticide.

The testimony brought out a familiar definition of safety and of effects, one widely shared by the medical profession. An exposure level that did not cause disease or detectable pathological problems among an exposed population was safe. Although Hayes admitted that animal experiments were our primary protection, "ultimate assurance about human safety must come from studies of people with prolonged and intensive exposure."[51]

Unlike its stand at the Delaney hearings, though, the Food and Drug Administration did not, in 1963, seek extensive changes in the structure of pesticide regulation. In 1950 it had asked for new laws that would give it authority to control the flood of chemicals entering the market; now it stressed the lack of an immediate hazard and the need for an extensive program of research to discover what long-range effects there might be. HEW Secretary Celebrezze said that the proper use of pesticides posed no immediate problems to human health, although animal tests did indicate a need for caution; the real problem, he believed, was whether continued use would contaminate the environment to such an extent that it would, ultimately, threaten both man and the environment. FDA

Commissioner George Larrick pointed out that the FDA was reexamining the tolerance levels for pesticides in the light of the new evidence, and that it would reduce any that should be changed.[52]

The Ribicoff hearings raised another issue—one that had not occupied Congress in 1950—the effect of pesticides on wildlife. Although Carson had been primarily concerned with human health, she had been strongly influenced by the death of wildlife in the massive spraying campaigns. As with human health, the committee found witnesses who stood on both sides. The agriculture department and economic entomologists were sure that pesticides were harmless. George Decker, for example, testified that there was not much harm done to wildlife in normal applications of pesticides. Eradication programs, he admitted, were somewhat different, but they were hardly the normal use of pesticides.[53] Wildlife biologists, on the other hand, were more concerned. Eldridge Hunt and James Keith, scientists with the California Fish and Game Commission, submitted their studies on pesticides in California. From their work at Clear Lake and in other aquatic ecosystems they concluded that "there is no predictable safe level of application when these toxic materials enter the food chains." Bioconcentration, their field work showed, was a danger to water birds feeding in any pesticide-contaminated environment.[54]

Representatives of the Department of the Interior, whose Fish and Wildlife Service had major responsibility for the environmental monitoring program, were less concerned than the Californians but equally certain that the situation required close attention. Secretary Udall called *Silent Spring* a "timely warning" that had raised "a searching series of questions."[55] He singled out the evidence on the transport of DDT and its bioconcentration as indications of potential problems and strongly urged more research. John Buckley, director of the Patuxent Wildlife Research Center, concurred. There was, he said, extensive spread of pesticides throughout the environment, but more evidence was needed. Buckley also pointed out one possible reason for the development of the problem; scientists had not had much interest in the problems of pesticide contamination in the environment.[56]

The hearings did not provide clear guidelines for legislation or administrative practice. The clearest message had been that the problems or potential problems of pesticides in the environment had been neglected, that neither foresight nor research had predicted the properties of the new chemicals once applied to field and forest. Still, the experts did not agree on what the evidence meant. With regard to human health and pesticides, two groups, with different experiences, professional training, and standards of safety had come to very different conclusions. One side had a negative verdict—that there was no danger from pesticide residues—whereas the other had a cloudy positive decision—there were indications of a real problem. The proof, though, was not conclusive; no official representative of a health agency stood ready to point with alarm, whereas some stood ready to point with pride to DDT's safety record. There was no question about the potential for harm to wildlife from pesticides that were misused; the problem was how dangerous they were if properly used. The massive fish kills in various rivers argued a need for caution but were not evidence that the new pesticides, properly applied, were dangerous. The studies of environmental contamination and bioconcentration suggested that discriminate use was impossible, at least in aquatic ecosystems, but were these effects being duplicated on a larger scale?

The committee faced, though, not only the ambiguous nature of the case, but the difficulty of getting accurate, adequate, testimony. Its report sharply criticized scientists who had been "inhibited by defense of past positions, employer loyalties, and lack of authority."[57] They had a special obligation to make their findings known to the public and to legislators and to make recommendations on the basis of their special knowledge. In this matter, however, the committee was running counter to scientists' ideas about proper conduct, ideas that had helped to stifle public discussion of pesticide residues in the environment before 1962 and that were to hinder the environmentalists in mustering open scientific support for a ban on DDT. Most scientists felt that they did not have, either individually or collectively, any responsibility for the application of their work or any obligation to use their expertise to inform the public, and they viewed with suspicion any of their number who did take a public position. Such work was a distraction from a scientist's primary responsibility—his work—and it tended to confuse science with politics. The physicists had to some extent lost this detachment after the dropping of the atomic bomb, but they were the only ones affected.

Peer pressure seconded personal desires, and in other cases superiors actively discouraged subordinates. Universities, particularly state-supported ones, tried not to alienate segments of the community that might provide support or political help, and a scientist speaking out ran the risk of rousing opposition within the institution itself, as at Wisconsin. State and federal scientists found it difficult, even if they were so inclined, to oppose DDT in public, for their funding was at the mercy of possibly hostile congressional or state committees. In come cases opposition was direct, as when one governor forbade officials of the state conservation department to testify in a lawsuit against the use of a persistent pesticide.[58]

The defenders of pesticides, on the other hand, came from institutions and disciplines that were accustomed to public advocacy and that, at least until the late 1960s, enjoyed general support for their position. Economic entomologists had a history of direct contact with the public and close connections through extension divisions at agricultural colleges. Lacking the inhibitions of their colleagues, accustomed to presenting their case and defending their work, and backed by powerful allies in industry and Congress, they mounted a strenuous campaign on behalf of persistent pesticides. Industry also had a carefully planned and, at least until the late 1960s, well-financed corps of spokesmen and scientists who defended the chemical.

Despite its problems with the scientific testimony, the Ribicoff Committee reached some decisions, though the report reflects the ambiguous nature of the evidence. It concluded that current pesticide use did not constitute a hazard to humans and that the present precautions were adequate to safeguard public health. Given the benefits of modern chemicals to agriculture and public health, it said, the risks inherent in their use were acceptable. However, it expected pesticide use to increase, and though it believed that the benefits and risks were not "*presently* [sic] unbalanced in any significant way," extensive research was needed to insure that new problems would not arise. It rejected sweeping claims about the safety of DDT. Although there was no evidence of health problems, there was, it said, insufficient evidence to rule out this possibility and serious doubt that current monitoring programs could deal with the hazards that might be posed by increased future use. As had the President's Committee, it called for extensive federal support for research on the effects of pesticide residues on all forms of life and more cooperation and a clearer division of responsibility among government agencies.[59]

Ribicoff hit on one of the key elements in the regulation controversy. "My own thinking," he said, "it is that present pesticide legislation and regulation is based on the premise that chemicals used for the control of pests remain where they are applied. . . . An increasing volume of evidence demonstrates, however, that the basic premise is false."[60] This went to the heart of the case against the use, as distinct from the misuse, of DDT. Economic entomologists could abandon mass spraying programs and concentrated sprays as misuse without harming their self-image or abandoning their authority; the deeper part of the case was that, given DDT's mobility and concentration through living organisms, there might be no possible safe use of the chemical. If Ribicoff was right, then the entire basis of pesticide legislation—its emphasis on immediate effects, concern with residues on crops, and its concentration in the hands of agricultural scientists—had to be changed. It was not a pleasant prospect.

The committee, though, had not found the evidence compelling enough to recommend, outright, sweeping changes, and both sides could live with the report. The only new legislation was the elimination of registration under protest, a move no agency had opposed, and a new law providing for a mandatory exchange of information on pesticide chemicals among the Agricultural Research Service, the Fish and Wildlife Service, and the Public Health Service. Manufacturers and pesticide users could take comfort in the lack of new controls on the use of pesticides and the failure to press for the rapid elimination of persistent chemicals. Conservationists also had something to take home; the committee had called for a full study of the effects of residues on man and in the environment. It looked as if pesticide policy would change, albeit slowly.

Part Three

CHANGING DIRECTIONS

CHAPTER 6

Moving toward Court

The ecological case against the chlorinated hydro-carbon insecticides as the pervasive factor in these phenomena [the population declines of several species of hawks] is essentially complete.

Joseph Hickey

May I strongly recommend that you take legal action against whatever agency plans to do the spraying. File suit to prevent the spraying.

Charles F. Wurster to Lorrie Otto[1]

NEITHER *Silent Spring* nor the subsequent public controversy over Carson's charges changed pesticide use and regulation in any significant way. Although the USDA curtailed the massive spraying campaigns that had caused so much public and scientific opposition, farmers and government agencies continued to use DDT. Even when it was replaced, its successors were not the nonchemical controls Carson had recommended—and which even the National Research Council had called the methods of choice—but other chemicals, sometimes more toxic than DDT. The apparent lack of progress, though, concealed several significant developments. Public opinion was changing, scientists were accumulating the evidence they needed to make the environmental case against DDT, and the basis of opposition to the chemical was shifting to emphasize the environmental effects of small doses rather than misuse. The opposition, frustrated at the failure of public education and lobbying, was also beginning to cast around for more effective ways to change policy. *Silent*

129

Spring, by making the debate over pesticide residues public, had decisively altered the situation; the period from 1963 to 1968 saw further drastic changes.

Everything ultimately depended upon the scientific case, for it was convincing evidence of DDT's mobility, persistence, bioconcentration, and effects on nontarget species that formed the environmentalists' case. Of all the research, perhaps the most important was the work done on the peregrine falcon, whose worldwide population crash after World War II Hickey termed "one of the most remarkable recent events in environmental biology."[2] Studies on the peregrine were a key element in both the Wisconsin hearing of 1968–1969 and the Environmental Protection Agency's hearing in 1971–1972 and had an important influence on the thinking of many biologists. The story is one of the most striking examples of the unsuspected effects of pesticide use on nontarget organisms far from the point of application and is an excellent illustration of the complex interdisciplinary research that went into assembling the case against the persistent pesticides.

The peregrine falcon, commonly called the duck hawk, is a slate-backed bird of prey about the size of a crow. It is cosmopolitan, found in environments ranging from the Arctic tundra to modern cities (where it nests on skyscrapers and feeds on pigeons), but uncommon. Subsisting at the top of food chains, it necessarily has a limited population, scattered thinly over vast territories, and, with the exception of the few urban birds, it is a wilderness dweller. Most people know only that it is reputed to be the fastest bird alive (supposed to reach 200 miles an hour when diving on its prey) and that it is used in falconry. Ecologists know it for other reasons—its nesting habits and normally stable population. Nests, called eyries, are reoccupied year after year by the same pair of birds, and favored sites are occupied by a new pair when the residents die. Records from Europe indicate that some eyries have been used continuously for centuries, and records in the United States date back to the 1860s. Because the peregrine depends upon several different sources of food its population is not subject to the fluctuations that some others undergo; indeed, "the general stability of peregrine populations—at least in the Temperate Zone—has been an impressive characteristic to ecologists."[3]

Around 1960, though, disturbing rumors spread about massive reproductive failure among peregrines and several other species of hawks. At the Twelfth International Ornithological Congress in the summer of 1962, Hickey heard a rumor that no peregrine in the northeastern United States had raised any young that year. Despite the already published evidence on the Florida eagles, he dismissed the story—there was no evidence. A year later, though, he read a paper that changed his mind—Derek Ratcliffe's report on the peregrine population of Great Britain.[4] Ratcliffe, an English naturalist, had undertaken a census of the peregrine for Nature Conservancy following a wave of complaints from pigeon raisers that a population explosion among the hawks was ruining pigeon racing. He found instead that the hawks were dying off. Field studies showed that the peregrine had begun to recover from its low point during World War II (when it had been shot to protect Army pigeons) but then had lost ground. Throughout the southern part of the British Isles, peregrines were not breeding, and even isolated northern populations were experiencing reproductive failure. The problem, Ratcliffe found, was closely linked in both time and place to the introduction and wide use of DDT. Citing Dewitt's work, he suggested that secondary poisoning was occurring, affecting reproduction.

Hickey was not the only one to notice the report; the editors of *Bird Study* and the Council of the British Trust for Ornithology had considered it so important that they rushed it into print ahead of other papers and wrote an editorial calling attention to the problem and stressing the need for haste.[5] Hickey was in a position to do something. If pesticides were causing reproductive failure in Great Britain, the same phenomenon should be taking place in the United States. Hickey decided to see if it was, and in the spring of 1964 he sent Daniel D. Berger and Charles R. Sindelar, Jr. on a 14,000 mile journey from Georgia to Nova Scotia, checking peregrine eyries. The trip, which followed the route of his own survey of 1939–1940, took the team to 133 known nesting sites, some of which were known to have been occupied for a hundred years. Meanwhile Hickey flew to Europe to confer with ornithologists there. They confirmed both Ratcliffe's report and the rumors of population collapses among several other raptors. Returning to the United States, Hickey saw the results of Berger and Sindelar's survey, which were even more startling. The pair had found not a single fledgling.[6] This, Hickey said, was "so dramatic as to make an international conference on the population biology of *Falco peregrinus* an absolute necessity."[7]

Hickey decided to organize a conference himself and, at the end of August 1965, sixty biologists, ecologists, and naturalists gathered in Madison. The meeting was not designed to solve the mystery—things were too complicated for that—but to bring together biologists from different areas to compare data, to contrast the situation of the peregrine with that of other raptors, to review "the implications . . . that these phenomena are associated with meterological, pesticidal, epizootic, or other ecological factors," and to discuss coordinating research.[8] Comparing data from two continents and from peregrine populations in several different ecosystems, the scientists quickly established the dimensions of the problem. There was indeed a worldwide population crash involving peregrines in diverse ecological and geographical areas. In some cases the population had failed to raise any young at all; reductions in the number of young by 80 or 90 percent were "not uncommon." The declines, they found, "followed a common pattern": first the peregrines laid infertile eggs and thin-shelled eggs that broke in the nest, then no eggs; one of the pair died, then the other, leaving the site abandoned.[9] It was also clear why the process had gone so far without alerting the scientists. Because reproductive failure rather than the death of the adults was involved, the decline was gradual, and only as the nonbreeding reserve was used up and eyries were abandoned did ornithologists become aware that something was happening. Even then the magnitude of the problem hid the cause. Egg collectors, pigeon fanciers—who shot falcons—disease, predation of eggs, or human disturbance of the nesting sites could account for individual population declines; only when the data was considered as a whole did these explanations fail.

Although the conferees could agree on what was happening, they could not agree on why. Pesticides were the obvious culprit, but were they guilty? Some, Hickey among them, were not convinced that the case against chlorinated hydrocarbon pesticides was strong enough to warrant the claim that they were responsible. There were just too many gaps in the evidence. In one discussion William H. and Lucille F. Stickel of the Patuxent Wildlife Research Center pointed out that there was no information available on the general environmental level of pesticides, the normal level in birds, or their average reproductive success in the wild. Eldridge G. Hunt of the California Department of Fish and Game, who had been involved in the Clear Lake studies, concurred. There was, he said, "insufficient data available to determine how serious the threat might be or even how pesticides might be affecting falcons."[10]

Others, though, were either convinced that pesticides were the cause of reproductive failures or thought the evidence strong enough to justify preliminary action. Tom Cade of the Syracuse University Department of Zoology pointed out a suspicious correlation between high pesticide levels and death and reproductive malfunctions. The Clear Lake studies, he said, indicated the possibilities of the situation, and he urged an "experimental" ban on the use of pesticides on the Atlantic coast.[11] Cade was not alone. The evidence, by now extensive and thoroughly checked, showed that population changes were too great to be passed off as coincidence, and even the skeptics agreed that pesticides were the most likely cause.

The "suspicious correlation" of pesticides and reproductive failure, coupled with the laboratory evidence, suggested several lines of investigation. The conferees called for studies to provide information on the normal level of pesticides and

their metabolites in various species and the distribution of
these compounds in the environment, experiments to find out
how pesticides caused reproductive failure (if they did at all),
and to see if the correlation between residue levels and
reproductive failure was one of cause and effect or if the
relationship was merely adventitious. They suggested research that would confirm or eliminate the major suspect, and
much of the work done in the next few years would spring
from the recommendations that came out of Madison.

The peregrine's problems were only the most dramatic of
the environmental events that attracted biologists' attention,
and the Madison conference was one of three held in 1965 on
the problems of pesticides in the environment. Late in May,
thirty scientists and naturalists met in Port Clinton, Ohio to
discuss the population trends of the bald eagle and possible
causes of its breeding failure. In July, eighty scientists from
Europe and America gathered under the sponsorship of the
North Atlantic Treaty Organization at Monks' Wood, Great
Britain to discuss the distribution and effects of pesticides in
the environment. There was some overlap at the meetings
but, taken together, the membership, which included scientists from government, universities, and industry, covered
the relevant disciplines and centers of study.[12]

Their discussions and recommendations were similar to
those at Madison. There were, the scientists agreed, serious
environmental disturbances; there was evidence that pesticide
residues were implicated, but there was not enough field data
or experimental work to satisfy most participants that chemicals could be blamed. Scientists clearly disagreed on the
significance of the evidence. Roland Clement, for example,
ranked low-level chemical poisoning first as a cause of eagle
mortality in the Great Lakes and Atlantic Coast populations
and urged Audubon to come out against the pesticides. His
colleague Alexander Sprunt IV argued that such a conclusion
was premature.[13] At both conferences, though, scientists
agreed on the steps that should be taken. What was needed
was more data on insecticides in the environment, better
analytical techniques to measure low residue levels, research

on the dynamics of insecticides in various ecosystems, and better coordination among scientific specialties and between scientists in different countries.[14]

The recommendations for research made it clear that the delayed expression of the pesticides (if they were the problem) had not been the only reason for the long delay in discovering the situation. The orientation of earlier research, concentrated on immediate problems and not on basic, long-term studies of chemicals in the environment, denied scientists the data they needed to assess the situation. Scientific specialization, necessary for research, proved to be a hindrance in this case. Since the problem did not fall neatly into one field, it was neglected, or scientists working on it saw only a single, narrow, aspect. Analytical chemists, ornithologists, physiologists, and ecologists were working on different parts of the problem without being aware of all the relevant research.

The need for improved methods of analysis for low levels of pesticides was overcome by the introduction of the gas chromatograph, which became a standard analytical tool by the mid-1960s. The new instrument, though, did not solve all the problems associated with the study of low levels of residues—indeed, it generated a controversy that dogged the scientific and legal battle over DDT through the final ban in 1972. The argument revolved around the analysts' ability to discriminate between DDT and a group of industrial compounds known as PCBs, which were also known to persist in the environment. To understand the basis of this argument and the environmentalists' case it is necessary to go a little deeeper into analytical chemistry. The gas chromatograph is, essentially, a tube filled with glass beads coated with a resin, equipped with an inlet and a stream of gas at one end and an outlet with a detector at the other. A sample, dissolved in an inert solvent, is introduced into one end, and the carrier gas sweeps it through the tube. The differing rates at which materials in the sample react with the resin provide the basis for separation. Nonreactive materials move through quickly, more reactive ones more slowly, as they bind to the resin,

undergo the reverse reaction that frees them, and move along
to undergo the same process farther down the tube. At the far
end a detector signals the presence of a chemical coming out
and measures the amount, and the instrument displays the
results on a graph of materials detected against time.

Since the detector does not actually identify the chemical
coming out, the basis for identification is a standard sample.
If, for instance, we know that, under a given set of experimental conditions (a temperature, a particular resin in the
column, and a given flow of carrier gas) we know that DDT
takes ten minutes from time of injection to pass through the
column, we may identify as DDT a material from an unknown emerging at ten minutes under the same conditions.
The instrument has the advantages of extreme sensitivity and
the ability to pick up all constituents in a sample, but these
characteristics have their own drawbacks. There can be a vast
number of "junk" signals that are of no use and merely clog
up the analysis. Two or more chemicals may have similar
retention times (the basis for the industry's attempts to cast
doubt on the analysis of the environmentalists' samples).
Common cross-checks include using different methods of
analysis to check the identity of the materials, running the
same sample under two different sets of experimental conditions (which may bring a separation), or reacting the sample
with something known to change some things but not others,
another way of changing retention time.

Despite the care needed in using the new instrument,
environmental scientists came to depend on it, for, properly
handled, it provided unparalled sensitivity and accuracy in the
detection of low-level residues. It was, though, only an aid in
the new program of environmental study, which involved far
more than advances in analytical chemistry. After the conferences of 1965, scientists set out to test the hypothesis that
pesticides were responsible for the environmental problems
they were finding, and in the next three years they gathered
new information, reassessed the significance of old, and fit the
pieces together into a coherent whole. It was an interdisciplinary effort, involving several lines of research—field studies

of the affected species, evidence on the distribution of pesticides in the environment and in food chains, physiological work on DDT's effects in mammals and birds, and laboratory studies growing out of the field work.

The peregrine falcon and other raptors were the first targets of the research, and one of the earliest investigations was of the egg-breaking phenomenon. How good was the correlation between the introduction of DDT and a general thinning of eggs in species now suffering from reproductive failure? In England, Ratcliffe undertook a crucial test of this hypothesis, assembling from private and museum collections throughout Great Britain a series of eggs laid since 1900. He found strong evidence of a correlation—a sharp drop in eggshell weight in 1947 and a significant difference in the mean weight of eggs collected before and after that date. Eggs from the years 1900 to 1946 had a mean shell weight of 3.81 grams, whereas those from 1947 to 1967 had a mean weight of 3.09 grams, an 18.9 percent decrease without a corresponding change in size. He found similar changes in sparrowhawk and golden eagle eggs from western Scotland (both birds had been reported as breaking their eggs) but no changes in ten other species that were not suffering population collapses. The work took on added significance when Hickey and Anderson found the same pattern in North American birds of prey. Postwar California peregrine eggshells showed an 18.8 percent decrease in weight compared with those of the period before DDT was in use. Again, there was an abrupt decline in shell weight in 1947. Other species showed the same characteristics. New England peregrines, East Coast ospreys, and Florida bald eagles all exhibited declines in shell weight in 1947 and lower shell weights thereafter.[15]

The environmental monitoring program, expanded in the wake of *Silent Spring* and the research of the early 1960s, shed more light on the fate of pesticides in the environment. Pesticide contamination, it showed, was high and spreading. In addition, the bioconcentration found in Clear Lake was being repeated in other bodies of water. Hickey, for example, studied the contamination of Lake Michigan ecosystems by sampling the lake on both sides of Door County Peninsula. Both Green Bay and Lake Michigan showed the same pattern of concentration through food chains that scientists had found in Clear Lake. Meantime, Wisconsin conservation officials were finding a similar pattern in inland lakes.[16] The study provided further evidence linking pesticides with reproductive failure. An investigation of herring gull eggs from Door County showed levels of DDE inversely proportional to the eggshell thickness. The probability of this relationship occurring by chance, Hickey said, "proved to be 1 in 1,000." The sampling program also revealed more about the degradation of pesticides in the environment and focused attention on one of DDT's metabolites, DDE. DDE was the product of one of the first stages of the degradation of DDT, but it was almost as resistant to further change as its parent. Stored in the fat, it was metabolized only very slowly. By the late 1960s the environmental monitoring program and tests of eggshells, birds, and embryos were pointing to DDE as the major factor in pesticide-induced reproductive failure.

Field studies, though, were only part of the evidence; physiological research suggested a mechanism by which the residues caused thin eggshells. The discovery was, in part, the result of an accident. In the early 1960s scientists investigating the metabolism of barbiturates in rats found certain animals reacted in a different fashion than others. They traced the anomalous results to the test animals' exposure to chlordane, a chlorinated hydrocarbon insecticide similar to DDT, and found that DDT, chlordane, and several other compounds accelerated the metabolism of barbiturates; the chemicals, they said, caused a proliferation of liver tissue that synthesized enzymes responsible for the degradation of steroids.[17] The result was a lower blood level of steroids as well as faster degradation of barbiturates. In 1967 David B. Peakall extended this work, showing that DDT and similar compounds gave the same effect in birds. This was a key finding, for the level of sex steroids in the female controlled the mobilization of calcium for the formation of eggshells. If the residues stimulated enzyme production, lowering the blood level of the sex steroid, they would cause a thin eggshell, just the effect observed in the wild.[18]

Wildlife biologists at Patuxent added another link to the chain. After the Madison conference, the Stickels had undertaken feeding studies to check the hypothesis that DDT caused eggshell thinning in raptors. They used American sparrowhawks, birds that were closely related to the peregrine but could be raised in captivity. Comparing the eggs of birds fed small amounts of DDT and dieldrin (another insecticide implicated in reproductive failure) with eggs from a control group on an insecticide-free diet, they showed that both chemicals caused thinning of shells and lowered reproductive success.[19]

The combination of studies convinced many biologists that DDT and its metabolites, particularly DDE, were the cause of widespread reproductive failure in the peregrine and other raptors at the end of residue-laden food chains. In April 1968, as he prepared the proceedings of the 1965 conference for the printer (even before the Stickels' work was complete), Hickey wrote that the ecological case was essentially complete.[20] By this time many other biologists had already come to the same conclusion.

The problem of the peregrine population declines illustrates the difficulty in tracing the environmental effects of DDT residues. No one was prepared for the astonishing effects of these compounds—their persistence, mobility, and bioconcentration—nor did anyone initially suspect that the most damaging effects would be on the endocrine system, not the central nervous system, where these substances had shown acute toxic effects. The delayed action and gradual decline had made it all the more difficult to connect the residues with reproductive failures. To the casual observer, and even to many ornithologists, there seemed to be no problem. As late as 1964, a Swiss ornithologist assured Hickey that there was no peregrine crisis; the birds were present in their usual numbers.[21] Only when repeated reproductive failure and direct death of adults had wiped out the nonbreeding reserve and eyries were abandoned did the

problem become apparent to all. The complicated nature of
the problem, its novelty—this was the first worldwide contamination by a synthetic chemical on anything like this
scale—and the necessity of interdisciplinary studies concealed
the problem from biologists.

Although reproductive failure among birds of prey was the
most dramatic evidence of DDT's unexpected side-effects in
the environment, it was not the only example. Scientists
working with fish had found that persistent pesticides, even
in very low concentrations, had a variety of effects: reproductive failure, increased mortality in fry, behavioral changes,
and increased susceptibility to stress. At the Fish and Wildlife
Services's Fish-Pesticide Laboratory in Columbia, Missouri,
scientists studied various species of trout and salmon, trying
to elucidate the effects of DDT in aquatic ecosystems. Scientists working for state fish and game commissions contributed their own data on these effects and on the concentrations
of chemicals in various lakes.[22] Scientists also studied other
areas and, in places as different as orchards and salt marshes,
found effects ranging from outright extermination of populations to shifts in the biological balance.[23]

Wildlife biologists were not the only people affected by the
scientific case; conservationists were increasingly troubled by
the growing evidence of environmental degradation. By the
mid-1960s the National Audubon Society was making pesticides a major target. In March 1965, in a speech to the
Society of American Foresters, Audubon president Carl W.
Buchheister pointed to this change. The Audubon Society, he
said, had begun in an attempt to save birds from extermination at the hands of plume-hunters. "Historically, there have
been two great broadenings of approach as we sought to cope
with conservation problems from decade to decade." The
first had come when the society, finding laws ineffective,
moved into conservation education and "took on the protection of a series of strategic areas essential either as nesting or
roosting or wintering habitats for wildlife." The second was
related to pesticides. At first concerned only with direct
damage, the Society had become convinced that a reappraisal

was necessary when Barker's 1958 study showed a DDT-earthworm-robin cycle. (Here Buchheister seemed to have been leaning on Clement's change of heart). Now contamination was general. Buchheister went on to cite the contamination of fish in the North Pacific, the discovery of residues in penguins on the Ross Ice Shelf in Antarctica, and recent work on birds of prey. What was needed, he concluded, were ecological solutions to insect-control problems and an orderly reduction in the use of persistent pesticides as recommended by the President's Science Advisory Committee.[24]

The shift from concern with "strategic areas" to a policy that aimed at the preservation of the complex web of relationships that made up the ecosystem marked a change in the conservation movement greater than that induced by the opening of public debate on pesticide residues. It marked the difference between conservation and environmentalism. The national park and bird sanctuary movement, which dated back to the nineteenth century, had had no quarrel with the industrial and technological development of America. It had asked only that some areas be set aside, preserved from the common lot of use for economic gain. Environmentalists, on the other hand, believed this strategy was useless in the face of chemicals that spread through the ecosystem. There could be no islands, no "strategic areas"; preservation of the wilderness, or the ecosystem, required more control over materials dumped into it. "Progress" must be subordinate to the interests of nature and man's need of nature. The environmental movement, unlike its predecessor, had direct implications for "progress" and American economic development. It called for changes not only in pesticide policy but in national economic policy, and at least implicit in the argument was much more public control over private actions.

Despite the growing case and the increasing efforts of the conservationists, there was no change in pesticide policy, and there seemed to be no redistribution of power—agricultural interests still had the only effective voice on the issue. To a certain extent, this apparent stability was an illusion, for as the events of 1968–1972 made clear, public opinion was slowly but surely shifting to the environmentalists' side. The defenders of persistent pesticides, though, had real advantages in bringing their case to the public and in affecting policy. The industry was quite willing to defend DDT, and the USDA, through its extension agents, had the ear of most of the farmers in the country. To go against the "massive promotional efforts of the industry and the propaganda of the Department of Agriculture" was, Audubon president Buchheister said in 1962, like "whistling into the teeth of a tornado."[25]

A significant advantage, particularly apparent in the early 1960s (though the industry relied on it through the last rounds of the fight), was the appearance of famous scientists with outstanding records. Wayland J. Hayes was the most prominent of these figures, but as late as 1972 others could be found to defend DDT. Norman Borlaug, a Nobel Prize winner, appeared to testify for DDT at the Environmental Protection Agency (EPA) hearings. More important in terms of policy was the industry's support in Congress. In the 1930s the FDA had contended with Clarence Cannon to get funds for studies of lead and arsenic residues; in the 1960s the foes of DDT had a similar battle with Jamie Whitten, a Congressman from Mississippi who held a key position on the House Agriculture Appropriations Subcommittee. Whitten vigorously defended the use of DDT—even writing a book, That We May Live, attacking Silent Spring—and he used his influence to slow or halt action unfavorable to its continued use. When the regulation of pesticides was transferred to the new Environmental Protection Agency in December 1970, Whitten retained his power; two months later the House gave his subcommittee control over environmental protection and consumer affairs.[26]

The industry also profited from the confusion of public debate, the lack of a clear-cut division of interests, a stark contrast between good and evil. Pesticides had immediate, visible, in some cases dramatic, benefits. Their drawbacks, on the other hand, generally appeared much later, were not obviously the result of pesticide use, and were not of immediate economic interest to anyone. Persistent pesticides, in short, had a natural constituency to whom they were very important; they had no enemies with a similar interest in suppressing them. Pesticides profited, too, from the general acceptance of science and technology; the public, at least into the early 1960s, was not inclined to question closely the miracles of modern science.

The industry's advantages left the environmentalists completely frustrated. Their case was growing, as scientists added new evidence, but their arguments seemed to carry no weight with the public, and the government seemed a closed corporation, deaf to their pleas and pressures. The only way out was to find some forum in which the defenders of DDT could not use their advantages of excellent public relations, impressive scientific figures, and powerful political support. The new way was found by a new organization, the Environmental Defense Fund (EDF), formed in 1967 to preserve the environment through legal action backed by scientific testimony. The formation of the EDF, with its tactics of confrontation and litigation, marked another change in the opposition to DDT. The early protests against the misuse of the chemical had shifted to opposition to its use, on the grounds that, once placed in the environment, it spread. Now the tactics shifted from public education (or propaganda) and lobbying to a direct legal challenge. The EDF was going outside the earlier arena, moving the contest out of the agencies, Congress, and the newspapers (though it was to use all these in some form) to the courts.

The EDF, now a national environmental group with offices in New York, Washington, Berkeley, and Denver, and with a budget of a million dollars a year, started with a small group of people on Long Island determined to do something about a local use of DDT. The original members were Victor J. Yannacone, Jr., a lawyer, and his wife Carol; Charles F. Wurster, Jr., a marine biologist at the State University of New York at Stony Brook; George M. Woodwell, an ecologist working at Brookhaven National Laboratory; Anthony S. Taromina of the New York Department of En-

vironmental Conservation; Robert E. Smolker, another
Stony Brook biologist; Dennis Puleston, head of Brookhaven's Information Division; Robert Burnap, conservationist, publicity man, and fund-raiser; Arthur Cooley, a high
school biology teacher, and (the only member off Long Island)
H. Lewis Batts, a Michigan scientist.[27] It grew out of the
interest, concerns, and observations of the members. Wurster, Taromina, and Woodwell were professionally and personally concerned with the environment. Cooley and Taormina had helped form the Brookhaven Town Natural
Resources Committee in late 1965 to preserve the natural
resources and wild areas around Brookhaven. Wurster had,
at about the same time, done a study of the effect of
Dutch elm disease sprays on robin populations in Hanover,
New Hampshire. Woodwell had seen the spruce budworm
spraying problem in Maine and New Brunswick and had,
somewhat warily, corresponded with Rachel Carson about
his observations. Yannacone had become involved with the
Brookhaven Town National Resources Committee and with
local fishermen in attempts to stop the degradation of the
Great South Bay—first by the pollution from the duck farms,
then through dredging operations.[28] A common thread was a
life-long concern with nature. Indeed, one member wrote "it
was a joke that to be hired by EDF in the early days you had
to be a bird-watcher."[29] By the time they formed the
EDF, the members were experienced in environmental education and scarred by several battles.

It was not the scientists, though, who started the legal
action; it was Carol Yannacone, and she did not do it out of a
scientific concern with the environment but because she
thought DDT was ruining Yaphank Lake, where she had
grown up. During the 1960s, she had become increasingly
disturbed about the fish kills there; in 1966 she had learned
that the kills were due to DDT and that the Suffolk County
Mosquito Control Commission was planning to dump in
another 60,000 gallons of DDT solution to kill the mosquito
larvae. She decided that something had to be done and
persuaded her husband Victor to take the case. In April he

filed a suit on behalf of Carol Yannacone and "all others
entitled to the full use and enjoyment" of the lake, asking for
an injunction forbidding the use of DDT by the county
mosquito control commission.[30] It was around this case,
Yannacone v. Dennison, that the group that formed the EDF
came together, and it was here that they tried out their
approach: environmental quality through legal action. The
Brookhaven Town National Resources Committee joined the
action, with Cooley, Woodwell, and Puleston working to
prepare the scientific case. Wurster read about it in the
newspapers and joined, eventually becoming the organization's scientific advisor and head of the DDT project.[31]

Yannacone v. Dennison came to trial in November, and the
presentation showed the preparation and tactics that were to
mark the EDF's approach: an assertion of the rights of each
citizen to a clean environment, an argument that the use of
persistent pesticides, because of their properties, was destructive of that environment, and a reliance on careful, detailed,
scientific testimony to prove its point. Speaking for the
group, Yannacone asserted that the suit was not one for
personal damages, nor was it to abate a nuisance; the entire
population of Suffolk County had an interest in a clean,
undegraded environment, and they were there to defend that
right. The scientific testimony centered on the properties that
made DDT dangerous when applied to the environment and
on the evidence of that damage. Yannacone used cross-
examination to show that the mosquito commission was not
well acquainted with that evidence. DDT, the group's scientists argued, was not necessary for mosquito control and its
use was dangerous to other forms of life. Using DDT in the
lake, Wurster argued, was like using atomic bombs against
street crime in New York City.[32] Finally, the scientists
reinforced their point with a technical appendix to Yannacone's legal brief—copies of the scientific papers that but-
tressed their position.

This case, like many of the EDF's, was a partial victory.
The group convinced the judge that DDT was harmful—he
maintained a temporary injunction against the mosquito

commission for a year—but he refused to ban its use. Only the legislature, he concluded, had the power to take such action.[33] Legally, the environmentalists had lost. In fact they had won. The commission, by now used to alternatives to DDT and fearful of more suits, continued to use other means of control. Yannacone and the scientists had managed to obtain what conservationists had struggled for for some years, a ban on a use of DDT. They had also discovered a new weapon against DDT: an alliance of legal tactics with scientific information in a forum outside the government agencies or Congress—a practical way to influence policy.

Everyone was enthusiastic and wanted to continue the partnership, taking on DDT in other places. There were two problems: setting up an organization and finding money to support the cases. The first was easily solved, though it took some time and discussion. In the spring of 1967 Woodwell sent around a scheme for an organization to be called the Institute for Ecological Jurisprudence; Wurster outlined plans for Conserve America; and Robert Burnap came up with something called the Environmental Defense Fund. Eventually, Yannacone recalled, they pooled ideas, taking Burnap's name and something from everyone's structure. Yannacone, the lawyer, was to be legal counsel, Woodwell, head of the scientists' advisory committee, and Dennis Puleston, head of the board of trustees. The rest of the group were the other members of the board and constituted EDF's total membership until April 1970, when it opened membership to the public.[34]

Money was harder to find, although there was considerable support for the new organization among conservationists. The Audubon Society had, with some reluctance, supported the mosquito case, finding funds for court transcripts, but it was still not ready to engage in direct legal action. In September 1967 Yannacone made a speech at the society's annual meeting in Atlantic City, calling for legal action to preserve the environment. The system of adversary litigation, he said, was the best method of redressing public wrongs, and he called on Audubon to support such action. H. Lewis Batts,

another member of the group, presented a resolution endors-
ing this plan at the members' meeting. It was "approved
enthusiastically."[35] The executive committee, though, hesi-
tated, and the EDF was incorporated without a formal
commitment from the Audubon Society. On 8 October the
group had an organizational meeting, and shortly afterward
Yannacone filed formal incorporation papers.[36]

The formation of the EDF was, in part, a response to the
frustrations of the founders, but it was also sparked by a sense
of urgency. By 1967 they were convinced that action must be
taken soon if irreparable damage to various ecosystems and to
several species of birds was to be avoided. The threat could
not be met simply by setting aside certain areas of the
world—DDT spread beyond the places where it was applied.
It had to be banned completely, and it had to be banned soon.
As a result, the EDF abandoned the genteel tradition of
conservation. Yannacone's motto was "Sue the Bastards!"
and, although the other trustees did not endorse the words,
they agreed with the sentiment. To Lorrie Otto, Wurster
wrote, "May I *strongly* recommend that you take legal action
against whatever agency plans to do the spraying. File suit to
prevent it." Public education, he wrote in another letter, was
"too slow and ineffective. . . . EDF represents an attempt to
do things another way; it is for those who have lost patience. . . . When it comes to changing an established order of
society, then I shall have to get action more rapidly than can
usually be achieved through conventional routes."[37]

The EDF's neglect of lobbying and public education and its
enthusiasm for confrontations with regulatory agencies distinguished it from other organizations. Its aims also set it
apart; the EDF wanted to do more than simply protect
wildlife. Woodwell said that EDF existed to "define human
rights through research, education, and litigation." There was
not enough scientific information to enable man to manage
the earth's ecosystems in a stable and sustained fashion, and it
was "the purpose of the Environmental Defense Fund, Inc.,
to provide a direct means of bringing science to bear on
environmental issues."[38] The other members agreed, though

there was, at least in retrospect, some division over the exact nature of EDF's mission. Yannacone recalled that the organization was designed as a temporary group to build a body of case law to establish citizens' rights to a clean environment and to educate the public through legal action. Wurster, in 1975, stressed much more the need to get results. He was much less concerned with public education and legal precedents, he said, than with getting action.[39]

The EDF began on shaky ground. Appealing as the idea of a citizen's right to a clean environment was, it was not well established in law, as Yannacone freely admitted. "We were," he said, "practicing a form of legal guerilla warfare. . . . There was no standing, there was no jurisdiction, there were no courses of action. . . ."[40] This did not particularly bother him, for he viewed the law as an instrument of social change, a way of resolving differences without appeal to force. "[I]f there is a social need that must be filled, and something must be done for the People, with a capital P, there must be a legal way to do it." It was the lawyer's job to find it, and if he could not find it, to "invent it."[41]

Yannacone was an excellent choice for the unorthodox task of bringing suit against a chemical on behalf of a hitherto undefined citizens' right to a clean environment. Besides his legal skill and dedication to the law as a means of settling issues, he had personal qualities that made him extremely valuable—aggressiveness, flamboyance, a quick mind, and the ability to grasp the essential points of an argument and shape them into a legal brief or an attack on an opponent. Scientists briefing him or being questioned by him were surprised at the speed with which he absorbed material and arguments.[42] He was neither daunted by the odds nor particularly bothered by the lack of precedents in this area. From his work with civil rights cases, he was accustomed to building a case rather than appealing to already established rules. Even his colorful courtroom manner was an advantage. The environmentalists needed, particularly in their first cases, to get public attention, both as a means of making their case outside the courtroom and of bringing in money, and Yannacone was always good copy for any reporter.

MOVING TOWARD COURT

149

The lack of a firm legal position did not stop the EDF from taking immediate action. One of the directors, H. Lewis Batts, was from Michigan, had done research on DDT and dieldrin, and wanted to challenge the pesticide programs in his state. He pledged $10,000. Ralph MacMullan, head of the Michigan Department of Natural Resources, also urged the EDF to come. It did, filing two suits. One challenged a cooperative state-federal insect control program—the Michigan and U. S. Departments of Agriculture planned to spray two and a half tons of dieldrin over part of Berrian County, Michigan, to exterminate an "infestation" of Japanese beetles. The other asked for an injunction against nine towns in western Michigan that were using DDT for Dutch elm disease control. Five days after celebrating its incorporation, amid mutual promises not "to rush into anything," the EDF had its hands full.[43]

It won a series of partial victories. On 24 October, 1967, when the dieldrin case came up, Judge Fox promptly threw the EDF out of the Michigan Court of Appeals "high, wide and handsome," ruling that under the Eleventh Amendment to the U. S. Constitution the organization had no standing. That afternoon, Yannacone got an emergency temporary injunction and went to the Michigan Supreme Court. It also denied the EDF relief, but the legal maneuvers delayed the program past the best spraying time, and it was put off for a year. The EDF was more successful against the Dutch elm disease control program. None of the towns contested the suit; all stipulated that they would not use DDT. EDF promptly filed another suit, naming fifty-six towns in the state using DDT. Within a year they all agreed to stop using it. Still, the EDF had not attained one of its objectives, the chance to present its full case in open court. It wanted to try DDT before the bar and before the court of public opinion.[44]

In October 1968, the EDF returned to the dieldrin action, bringing suit in United States District Court for the Eastern District of Wisconsin.[45] This was environmental law with a vengeance. To establish jurisdiction EDF had to show that the state of Wisconsin had a vital interest in a "regional ecosystem," Lake Michigan, and that the use of dieldrin in Berrian

County, Michigan would cause damage to that ecosystem. The court record shows just why the environmentalists found litigation a useful, but not a perfect or easily applied, remedy.

Michigan vigorously supported the use of persistent pesticides. In 1967 the governor had forbidden the employees of the Department of Natural Resources, including MacMullan, to testify for the EDF, and in 1968 the assistant attorney general, representing the state department of agriculture, disputed the EDF's standing, the court's jurisdiction, and the charge.[46] The case, he contended, was the same one that had been thrown out of the Michigan courts, and it could not be brought in Wisconsin. The action did not arise in Wisconsin, nor did the plaintiffs live there. Their charge was also defective; the EDF lacked an essential party to the suit—the Secretary of the USDA. The attorney general also claimed that the action was causing serious economic damage. As long as the nurserymen of Michigan had to contend with the Japanese beetle, their business was hampered by federal quarantine regulations.[47]

Yannacone replied that there was concurrent jurisdiction. Both states had an interest in the lake, which would suffer "permanent and irreparable damage" from the proposed sprays. As for the damage to nurserymen, this was "nonsense." New York nurserymen had managed to carry on their business for years under these regulations.[48] His basic argument, though, was that the EDF did not get a fair trial in Michigan. It had had only one day to present its evidence, and this had caused a severe contraction of the scientific testimony. The justice of a "day in court," he said, was not met by a literal day. The plaintiffs needed a chance to present all the evidence.[49]

The judge did not, at least, throw the EDF out; he allowed it to proceed with the case. It opened with a strong attack on the wisdom and necessity of the dieldrin program. Yannacone put B. Dale Ball, director of the Michigan Department of Agriculture, on the stand and asked him about the criteria his department used to define an infestation. It was planning to treat 4,800 acres because it had found a few beetles in traps.

How many beetles made an "infestation"? One? Ball was extremely evasive on this point, and Yannacone finally moved to strike his testimony on the grounds that Ball did not understand the concept. Even if there was an infestation, he went on, was eradication the best means of coping with it? The department had already eradicated the beetle from the area around Battle Creek several times. Since there were recurrent infestations, was it worth the cost and the damage to wildlife? What evidence was there that the beetle would cause any loss to Michigan's agriculture?[50]

The EDF then presented witnesses to show that the program would cause damage to the ecosystem of the lake. Ralph MacMullan, director of the Michigan Department of Natural Resources, said that the proposed sprays would be dangerous to wildlife of the area; dieldrin was "one of the larger, more important [pollution] problems" threatening the state. It was persistent, mobile, and broadly toxic. He strongly supported the EDF's original argument; Lake Michigan was, he said, a regional ecosystem, and any dieldrin that entered the lake was of concern to both states.[51] Three more witnesses testified on specific parts of the argument. Howard D. Johnson, assistant professor of fisheries and wildlife at Michigan State University, said that the lake was already polluted. DDT residues in Lake Michigan fish were two to five times those in Lake Superior specimens, and there were measurable dieldrin residues in the former. Charles T. Black, a biologist, testified that if dieldrin were applied to the Berrian County lakeshore, as the Department of Agriculture planned to do, it would move into the atmosphere and the lake, causing damge to the ecosystem. Finally Wurster gave evidence on the toxicity of dieldrin and DDT to various phyla and on the properties that made these substances so dangerous to the ecosystem—broad biological activity, mobility, chemical stability, and bioconcentration through food chains.[52]

In the end, though, the EDF got little more than it had received in Michigan. The judge rejected their arguments, cut their presentation short, and allowed the spraying to proceed.[53] The environmentalists turned again to DDT and

Dutch elm disease control, the subject that had brought them to Wisconsin, and here they found their forum against persistent pesticides. The new case was the result of the efforts of local conservationists to halt the use of DDT. In the wake of *Silent Spring*, a group of Wisconsin conservationists, organized into the Citizens' Natural Resources Association of Wisconsin, Inc., (CNRA) had actively campaigned against persistent pesticides, holding seminars to educate the public and testifying against the use of DDT for Dutch elm disease. The group had some success; a few towns in the state stopped using DDT, but progress was not rapid.[54] In August, 1968, Lorrie Otto found a notice in the local paper: the Wisconsin Department of Agriculture had again recommended DDT for Dutch elm disease control, and the towns around her, and Bayside as well, were planning to use it. She called Joseph Hickey, who agreed that the only thing left was a suit. A week later, when they attended a National Nature Conservancy meeting on Long Island, she and Hickey went to see Wurster, with whom she had been corresponding for over a year.[55]

The meeting was useful to both parties. Hickey and Otto needed legal help if they were to fight the sprays; the EDF needed a good case if it was to survive. What made Otto's case so attractive was the information she had been collecting on DDT use in Milwaukee County. Wurster promised that the EDF would come to Wisconsin and represent the CNRA in a lawsuit against DDT use. He asked Otto to find a lawyer to introduce Yannacone at the bar, to start a campaign to raise at least $15,000, and to tell reporters the story.[56] The EDF wanted, and needed, a lot of publicity. The CNRA subsequently filed a complaint with the Wisconsin Department of Natural Resources (DNR), which was reponsible for oversight of these programs, naming as defendants the city of Milwaukee and Buckley Tree Service, which it alleged had a contract with the city to spray the elms. Maurice Van Susteren, chief hearing examiner for the department, assigned himself to the case and set the hearing for 18 October.[57]

The hearing, held a week after the dieldrin case, was a fiasco. It began with a two and a half hour off-the-record discussion that effectively closed the case. The city and Buckley agreed not to use DDT and showed that there was no contract between them. (They further agreed not to take any action against the plaintiffs—specifically a countersuit.) Yannacone then asked for a three-day adjournment to prepare an amended complaint, and a six-day hearing in which to present the full case; Buckley of Buckley Tree Service had agreed to act as the defendant in the action. Van Susteren, though, said that this was impossible. The hearing schedule was full for the next two months, and, the complaint was, with the stipulations of Buckley and Milwaukee, moot.[58]

It seemed that all the environmentalists' efforts had been wasted, and some of them, including Lorrie Otto, began to cry. Van Susteren remarked to the clerk that these people were crazy. They had won, what did they have to be unhappy about? What, he asked Yannacone, was the matter? The lawyer explained that their goal was not so much to ban DDT in any particular place as to find a public forum and an impartial arbiter before whom to present their scientific evidence and get a judgment. Van Susteren pointed out that if that was the case, they had chosen the wrong legal route. Wisconsin law provided a declaratory ruling procedure that was much more suitable. Any Wisconsin citizen could ask a government department for a ruling on the applicability of a particular set of facts to any rule enforced by the department. The concerned agency would then hold a public hearing in which each side in the dispute would present its case, subject to rules of evidence and cross-examination.[59] CNRA members, for example, could ask the Department of Natural Resources if DDT was a water pollutant under the Wisconsin water-quality standards. The department would then hold a hearing in which the EDF could attempt to show that DDT did, in fact, contaminate the waters of the state and injure the human or animal life of the area.[60]

This was ideal. A hearing on a declaratory ruling, designed

to allow the clarification of administrative rules and procedures without a justiciable matter, would allow the EDF to
concentrate on its scientific case, rather than any particular use
of the insecticide. It could sue DDT. It lost no time doing so.
On 28 October Frederick L. Ott, on behalf of the CNRA,
asked the Department of Natural Resources for a declaratory
ruling: was or was not DDT a water pollutant? Van Susteren
assigned himself to the action and arranged for a ten-day
hearing, to begin 2 December in Madison. The EDF had its
case.

CHAPTER 7

A Legal Tour of
Round River

> One of the marvels of early Wisconsin was the
> Round River, a river that flowed endlessly into
> itself, and thus sped around and around in a never
> ending circuit. Paul Bunyan discovered it, and the
> Bunyan saga tells how he floated many a log
> down its restless waters.
>
> No one has suspected Paul of speaking in
> parables, yet in this instance he did. Wisconsin
> not only *had* a round river, Wisconsin *is* one.
>
> Aldo Leopold[1]

THE PROSPECT of an extended hearing was exhilarating, but
it was more than the EDF had bargained for; it had neither the
money nor the scientific witnesses for such an undertaking.
Very quickly, though, one of the advantages of legal action
became apparent—people were far more willing to support an
action that promised concrete results than to help in efforts to
educate the public. Local environmentalists quickly rose to
the challenge. Fred Ott, a wealthy Milwaukee bird lover,
took charge of raising money in that city, and Orie Loucks, a
botany professor at the University of Wisconsin, Madison
(and a CNRA member), began work in the capital. By the
time the hearing opened, the CNRA had raised $25,000.
Volunteers offered housing and transportation for witnesses
and for Yannacone and Wurster; Mrs. James F. Crow, wife of
one of the panelists from the 1963 discussion ("Insecticides
and People"), provided a central mailing and information
service. Others rented an office and organized the typing

155

service that transcribed the hearings each day from Yan-nacone's tape recording.[2]

Local scientists were just as helpful; faculty and graduate students at the University of Wisconsin, Madison, arranged a scientific reference service for the EDF. A week before the hearing began Hickey asked William Reeder, a member of the zoology department, to help prepare information on the scientific background for the environmentalists. Reeder, initially detached, was soon caught up in the enthusiasm and recruited his own and other graduate students (including some in economic entomology), to do the library work. During the hearing he had a place at the counsel table, and whenever someone mentioned a paper, quoted a scientist, or brought up a new line of questioning, he would send a messenger out to call his assistants at the university. They in turn would look up the article, find out the scientist's background and qualifications, or assemble a group of papers on the topic under discussion. Usually Yannacone had a fund of information within an hour.[3]

Reeder's service, though, was only designed to provide information on topics that came up during the hearing; the burden of preparing the EDF's case fell on Wurster, now head of the Scientists' Advisory Committee and chief project officer for DDT (Woodell had resigned in February 1968, and Wurster had taken over the committee. Until the EDF expanded in the early 1970s, he remained the major scientific support of the organization and continued to direct the DDT case through the final ban). Although he was less conspicuous than Yannacone during the hearing, Wurster was just as essential. It was he who organized the EDF's argument, found scientists to testify on each point, convinced them to come, and served, with Yannacone, as the EDF's staff in Madison during the hearing. In addition he did much of the detail work, solved day-to-day problems, settled arguments, and soothed ill feelings.

Wurster had to plan a case as if for trial, for the hearing resembled one. The EDF, representing the petitioners—the CNRA—would present the direct case. Then those opposing the petition, the intervenors, would present a rebuttal, followed by a redirect case by the petitioners. The hearing was governed by basic rules of procedure and evidence; witnesses had to be qualified to present expert testimony; they were under oath and were subject to cross-examination. On the other hand, the examiner had a great deal more leeway than a trial judge in deciding what was relevant to the issue at hand. There was no concrete issue, no particular use of DDT at stake; the hearing was, legally, to decide if, according to the water pollution laws of Wisconsin, DDT was a pollutant. Finally, other parties could enter the hearing without committing themselves to either side. The rules provided for appearances "as interests may appear," with witnesses presenting evidence they felt bore on the case.

Wurster planned to use five of the ten days, 2 through 6 December, to present the evidence. The EDF would begin with a description of the natural resources and values affected by DDT, move to the usage of and controls on the chemical, then discuss its properties, and end (the last three days) with the environmental damage it caused. The first witnesses would be political figures—including Senator Gaylord Nelson. Agricultural scientists would follow to describe DDT usage and controls, and then a variety of scientists would establish DDT's environmental properties. By late November Wurster had outlined the case and found witnesses for most of the parts. They included Wurster himself (to summarize the EDF's contentions), George Wallace from Michigan State, several fisheries biologists from Michigan (some of whom had been involved in the dieldrin action), Joseph Hickey, David Peakall, Robert Risebrough, Robert Rudd, and George Woodwell. Robert van den Bosch, an entomologist working on biological controls, was scheduled to discuss his specialty on 6 December, and another witness would end with a discussion of the functioning of federal and state control policies.[4]

The progress of the hearing made hash out of this neat plan. Delays, cross-examination, and new points stretched the hearing long past its ten-day limit; schedules made it impossible to present witnesses in correct order, and the intervenors' witnesses, brought by the NACA, caused the EDF to introduce new ideas, evidence, and witnesses in its rebuttal. What emerged was a case with five major points. The EDF emphasized, first, that there was an ecosystem—an interrelated system of air, water, soil, and biological organisms that lived in the system. Second, DDT, unavoidably—because of its physical and chemical properties—contaminated the ecosystem when released in it at any point. Third, residues of the chemical had adverse effects of wildlife. Fourth, they were a potential hazard to humans; the available evidence did not show that DDT was safe. Finally, it was possible, the EDF argued, to replace DDT with better, safer means of control.

The different parts of the argument were designed to meet various problems the EDF faced. The first point—the easiest to prove—was simply the basic ecology course the EDF used as background. The second, the properties of DDT, was well established, but crucial. The EDF was contending that DDT was a danger because, applied anywhere, it spread through the ecosystem. The evidence of damage to wildlife, the third point, was the heart of Wurster's planned case. It was essential for the hearing and, to a lesser extent, for the national campaign against DDT. The last two points were part of the EDF's attempt to discredit the industry's defense of DDT, particularly for the national effort. The industry's basic defense was that DDT was safe for man and essential for agriculture and public health. Only if the EDF could show that it was not proven safe, if it could cast doubt on the industry's and the Public Health Service's assertions and present a workable alternative, could it hope to win the battle for public opinion.

The planning of the case shows another of the advantages of quasi-judicial proceedings over efforts at public education. In the hearing room the EDF could define the issue—was DDT safe for man and wildlife? Although it and the intervenors could raise other issues that went outside the legal case, such as the adequacy of federal oversight and the possibility of replacing DDT, the main thrust of the hearing was the issue the EDF wanted to discuss. The hearing, in effect, allowed it to choose its ground, set the terms of discussion, and force the industry to come and defend itself.

The industry remained unaware almost to the last of the danger. The Industry Task Force for DDT of the National Agricultural Chemicals Association, the industry lobby, did not even find out about the hearing until the day before Thanksgiving when one of its members, Louis A. McLean, read in the paper that there was going to be a hearing on DDT in Wisconsin the next week. McLean, a retired lawyer and former counsel for Velsicol Chemical Corporation, had been an ardent defender of DDT for years and was well prepared to present a case for the chemical to a legislative committee (he assumed that this would be the format of the Wisconsin hearing). He called the NACA's office in Washington and suggested that he go to Madison, a few hours drive from his home near Chicago, and present the case.[5]

DDT's defenders paid dearly for their lack of preparation. In Madison, McLean appeared as a fumbling, ineffective advocate, not because he was stupid, but because he was pitched into a situation for which he was not prepared. He was an experienced lobbyist with a well-deserved reputation for his skillful presentation to legislative committees,[6] but he was not a trial lawyer, and by the time the NACA provided effective legal help, in January, much of the damage had been done. The environmentalists had the essentials of their case in the record, including the most controversial points, had established the trend of the hearing, and had completely dominated the early sessions. The fault, though, should not be laid at McLean's door. The industry failed to appreciate the symbolic nature of the hearing, the psychological impact of the EDF case in the hearing room, or the effect of a successful "prosecution" of one persistent pesticide on the cases against others. It never gave McLean the scientific backing he needed to make a respectable case or a good effort in cross-examination. One industry representative later admitted that he had been ordered to support the effort only "as long as it did not cost a penny."[7]

On 2 December, Maurice Van Susteren opened the hearing
in the assembly chambers in the state capitol in Madison. It
quickly became apparent that the hearing would be a public
and political, as well as a scientific and legal, case. The first
witness was Senator Gaylord Nelson, who took the stand to
talk about his own increasing concern about DDT. He
stressed the importance of the Wisconsin hearing as the first
step in a nationwide ban. "Dangerous environmental contamination is occurring at a rapid and accelerating pace," he
said, and "[t]his hearing affords an opportunity to take a
significant step that may well have historic consequences."
(18)[8] His testimony had nothing to do with the scientific case
against DDT, but it did provide a dramatic opening and
attracted public attention. The (Madison) *Capital-Times* carried a banner headline in red ink over the masthead: "Nelson
Urges Ban on DDT." The story said that Nelson had given a
"stern warning" about the "environmental disaster" coming
from DDT pollution and had cited "chilling evidence" of
DDT's long-lasting effects on wildlife.[9]

There had clearly been a change in public opinion since
Silent Spring. In 1962 "environment" had been only a strange
word; in 1968 it was a political issue. There had been
Congressmen on both sides of the debate over *Silent Spring*
and much sentiment backing the "farmers" against the "bird
watchers." In 1968 there was much less open political support
for the continued use of persistent pesticides and little political
capital to be made from defending the manufacturers and
users of these chemicals. Testimony by Nelson and State
Assemblyman Lewis Mitness was only one aspect of the
political interest in the question. The Wisconsin public intervenor, Robert McConnell, attempted to enter an appearance
"as interests may appear." Van Susteren ruled that McConnell could not appear unless requested by the agency holding
the hearing, the Department of Natural Resources. McConnell promptly went to Dane County District Court and
secured a writ of mandamus from Judge Norris Maloney. He
argued that Van Susteren's action denied the public intervenor's authority to participate and represent the public in

matters that affected the state, and Maloney agreed. So the public intervenor came, representing the citizens of Wisconsin, and ultimately played a small, but useful, part in the presentation of the evidence.[10] These maneuvers reflected the political climate of Wisconsin—liberal and with a strong environmental movement—and the desire of the public intervenor to take a hand in the proceedings, but it was also more evidence of the shift that had taken place since 1962.

With the political witnesses out of the way, the EDF moved to establish its opponents' case, to give it "something to shoot at." It called, adversely, the counsel in opposition to the petition, Louis A. McLean. After a few questions on McLean's background, primarily to establish his connection with the chemical industry, Yannacone produced an article—the transcript of a speech McLean had published the year before in a semipopular scientific magazine, *Bioscience*. Reading from the speech, Yannacone established that McLean believed that people who were opposed to the use of pesticides were either misinformed or mentally unbalanced. They were, McLean had said, people driven by obscure sexual urges, "first described by Freud," and were usually cranks. "The antipesticide leader . . . can almost always be identified by the numerous variant views he holds about regular foods, chlorination and floridation of water, vaccination, public health programs, food additives, medicine, science, and the business community." The leaders in this movement used the controversy to sell "natural foods at unnatural prices, [and] to give color to their books, writings, and statements to gain notoriety. . . ."[11] They were antisocial and opposed to progress. Their opposition to pesticides had caused the death of millions through malnutrition and disease, millions who might have been saved by the use of agricultural chemicals. Did McLean, Yannacone asked, really believe this? The exchange was inconclusive, but the incident was more than an attack on McLean and more than an attempt to get publicity. The EDF wanted to establish that the intervenors did not take the environmentalists seriously, indeed regarded them as unworthy of an answer, and to show the public the type of attack they had endured for years. The point was particularly important in view of the petitioners' case, which the EDF presented with great stress on the thoroughness of the scientific studies. McLean's article was a sharp contrast to the careful, factual case presented by the petitioners, and the EDF wanted to emphasize the difference.

McLean's appearance on the stand gave the first demonstration of another advantage of the hearing over public debate—the chance to question the defenders of DDT on their ideas and statements. The EDF's rough treatment of McLean served notice that the petitioners intended to use the chance to challenge arguments the defenders of DDT had used for years. It became particularly evident during the cross-examination of the intervenors' witnesses, when Yannacone unmercifully attacked the industry's experts. Cross-examination brought about a close cooperation between Yannacone and the scientists. They briefed and coached him in the various scientific specialties, making sure he could challenge each witness. Yannacone, on the other hand, carefully prepared each EDF witness before he took the stand, impressing on him that the hearing was not a scientific forum, but an adversary proceeding in which a cross-examiner would attempt to discredit his testimony and research. One scientist said that "after Vic was through with me that night I wasn't sure about my own name, let alone the validity of my research."[12] The work paid off. The petitioners' witnesses were well-grounded and cautious, and the intervenors did not seriously discredit any of them, whereas the pro-DDT witnesses came under strong attack for which they were not prepared.

To establish the ecological part of its argument, the EDF called Hugh H. Iltis and Orie Loucks, both professors of botany at the University of Wisconsin, Madison. Yannacone had told Iltis to "lay it on with a shovel," but not to mention DDT. He obliged. Some of his testimony about the native flowers of Wisconsin (illustrated by slides) was quite poetic, and the examiner had to remind him gently that he was entering into "argument." Wisconsin, Iltis testified, formed a regional ecosystem, intimately connected with the rest of the country. Within the area animals and plants formed an interdependent whole. Some plant species, for example, depended heavily on insect pollination; without the insects they could not reproduce. Many of the native wild flowers, he said, might disappear if continued pesticide use reduced the species they needed.[13]

Between Iltis's and Loucks's appearances the EDF brought in a witness on the legal procedures for state and national registration of pesticides, part of its plan to challenge the current regulatory system. He was Ellsworth H. Fisher, professor of entomology and coordinator of pesticide-use education in the extension division of the University of Wisconsin, Madison. Fisher, whom an unfriendly journalist once labeled the "epitome of the squirt-gun entomologists," was reluctant to appear.[14] He testified, in fact, only under threat of a subpoena, and was so sympathetic to the inter- venors and so unsatisfactory a witness that Yannacone had him declared a hostile witness so that he could cross-examine him.[15] Despite this, Yannacone managed to make the EDF's point, that the agencies in charge of registering pesticides relied on manufacturers' studies of effectiveness and safety, and confined their tests to studies of acute poisoning. They did not require any tests of chronic effects of pesticide residues on man and wildlife. Yannacone also brought out the effec- tive isolation of the Department of Agriculture. There was no way for an agency or for the public to initiate a review of the registration of a pesticide; it had to be done by, and within, the Department of Agriculture.(107–160)

This raised the question of whether federal regulation of pesticides was adequate. DDT manufacturers claimed that government studies of pesticides and regulation were entirely adequate to protect the public and the environment. The petitioners were trying to show that federal regulation was pro forma and did not even attempt to consider the dangers of environmental contamination or the biochemical effects of pesticide residues on man. Regulation was inadequate because it dealt only with gross pathological effects on man and acute

poisoning of wildlife, and because the Department of Agriculture did not update and reconsider registration in the light of
new scientific information.

After calling Loucks to expand on Iltis's testimony about
the properties of the regional ecosystem, the EDF proceeded
to the main part of its case. On 3 December, Yannacone called
to the stand Charles F. Wurster, the EDF's expert on pesticide
contamination in the environment. Wurster outlined the
petitioners' case against DDT and presented specific examples
of the environmental problems it caused. He began by
describing the chemical and physical properties that made it
so dangerous. DDT, he said, had broad biological activity; it
affected not only the target insect but other insects, birds,
fish, mammals. In high concentrations it was a central
nervous system poison, inducing tremors, lack of coordination, and, ultimately, death. Even in minute doses it upset
enzyme balances, causing reproductive problems. It was also
extremely stable under environmental conditions; placed in
the soil DDT had a half-life of more than a decade. Despite a
very low solubility in water (1.2 parts per *billion*) it spread
easily through evaporation, codistillation, and mechanical
transport of material to which it was sorbed.[16] The combination of mobility and persistence meant that DDT placed
anywhere in an ecosystem eventually permeated the system.
Finally, DDT had a high solubility in lipid tissue; animals in a
contaminated area stored the chemical in their tissues in much
higher than environmental concentrations, and low levels in
the environment could become sufficiently concentrated
through natural food chains to poison species at the higher
end of the chain.(326–340) His argument was not that DDT
was causing pathological changes. It was upsetting the estrogen level, interfering with the mobilization of calcium for
eggshell formation. The birds produced eggs with thin shells,
which broke in the nest, or eggs with such high concentrations that they failed to hatch. It was not necessary that the
residues kill the adults. "It is possible," he said, "to cause the
complete collapse of a population and the extinction of a
species without ever killing a single individual."(450)

Wurster was the first witness to develop a significant and controversial part of the petitioners' argument, and McLean spent a day and a half trying to discredit him. The cross-examination, though, only showed how ill-prepared McLean was to question the environmental evidence against the chemical. Without scientific advice, he could only attack the witness and the EDF. He began by trying to show that the Environmental Defense Fund was a crank organization, dedicated to banning all pesticides regardless of their properties and of the consequences to public health and food supplies. When he later returned to this line of questioning the examiner asked about the relevance. McLean said that:

> I think it will become apparent, Your Honor, that while this particular action is limited to DDT, that the activities of this organization in the last few years will indicate that they are not simply against one chemical, but any, the use of any pesticides.

Yannacone: That's a pile of hogwash.(478)

McLean also attempted to show that Wurster was one of the "bird watchers," a man with no practical experience in farming, forestry, or public health, completely ignorant of the many benefits derived from using DDT. Was Wurster an entomologist? Had he done work with DDT to control insects or to increase crop production? Had he lived on a farm? Had he done any research on humans? It was the familiar charge that opponents of DDT were more concerned with birds than with human lives or crops, and the familiar argument that the only people qualified to assess pesticides were those who had applied them in agriculture or public health programs.[17]

McLean's cross-examination brought out differences in the way in which the two sides approached the problem of scientific qualifications. The EDF's scientists believed that the problem of residues in the environment cut across disciplinary boundaries, and they advocated a multidisciplinary approach. Since there were no formal qualifications and program of study for the new environmental science, they tended to think in terms of the design and execution of experiments. The industry, on the other hand, was concerned with formal qualifications in defined academic, scientific specialties. Qualifications, for them, meant a degree in a particular field of research, membership in professional organizations, and the approval of scientific peers. They attempted to limit the competence of a witness to the discipline in which he had taken his degree, whereas the EDF tried to show that the intervenors' witnesses had neglected important parts of the problem by limiting the scope of their investigations and not planning their work in a larger context. The EDF did not ignore qualifications in the conventional sense—indeed it took pains to show that its experts were as respected and prestigious in their fields as those introduced by the intervenors— but it thought of academic specialties as tools, not as watertight compartments.

Wurster's testimony was vivid and dramatic and the newspapers seized on it. The *Milwaukee Journal* headed its story: "DDT Uncontrollable, Scientist Says," and in bold-face type quoted Wurster: "The more you use it, the more you need to use it. We have created a more serious pest problem than we ever had before."[18] The stories, and the headlines, were evidence that the petitioners were bringing their case to the public. Part of the publicity was due to EDF's policy of friendly relations with the press; it made every effort to give reporters a story and make sure they got it right. It could not have had the sustained coverage it did, though, unless it had a story and convinced the papers it was worth covering. For example, the *Capital-Times*, Madison's afternoon paper, had decided that the story would not be worth sending a reporter each day. Whitney Gould, assigned to the hearing for the first few days, argued with her superiors that she should remain. They finally agreed and the *Capital-Times* had a story every day, bringing the hearing to the central part of the state. The Milwaukee papers also gave good coverage.

After Wurster, the EDF brought to the stand Robert van den Bosch, an entomologist at the University of California,

Berkeley, who was responsible for recommending pesticide spraying schedules for California field crops. He had been scheduled to discuss nonchemical controls at the end of the EDF's presentation—Wurster had assumed that the direct case would be almost finished by 6 December—but McLean's long cross-examination put van den Bosch in at the beginning of the presentation. He gave an impressive account of the problems DDT caused, discussing the chemical's effect on natural systems of control, and showed how entomologists were working to replace exclusively chemical controls with a system of integrated controls, which would take advantage of natural enemies and cultural practices and rely on chemicals as little as possible. His most important contribution, though, was to undermine the distinction McLean had tried to make between the "practical" men and the "bird watchers." Van den Bosch had recommended DDT for years and had closely observed its effects in agricultural ecosystems before deciding that it should not be used. He also showed that entomologists were not united on the virtues of DDT, a point that was easily overlooked. Even during the height of the enthusiasm for DDT there were entomologists who warned against an over-reliance on chemicals, but their votes had been drowned in the general clamor for the chemical. In 1951, for example, a Canadian entomologist, G. C. Ullyet had warned that the economic entomologists had to be "thoroughly conversant with the principles, methods, and aims of ecology." Concentrating on chemical control, "he ceases to be a biologist and . . . becomes a mere tester of poisons or an insecticide salesman."[19] Others had repeated these warnings.[20] Clarence Cottam, a U.S. government biologist, sought for many years to convince his fellow scientists that chemical control of insect pests should be replaced with ecological controls, which would stabilize pest populations at a low level.[21]

Van den Bosch brought these points of view into the open. He also made a big impression on the newspapers. The *Milwaukee Journal* headed its story "Scientific 'Convert' Warns State on DDT." The *Capital-Times* said "Ex-DDT Backer Says He Wouldn't Approve it Now" and cited in boldface van den Bosch's statements that entomologists had "created a monster" through the use of DDT, and that they had seized on it while "totally ignorant of the genetic and ecological implications of its use."[22]

The next witness was Robert Risebrough, a molecular biologist from the University of California, Berkeley, who testified on two subjects, the biochemical mechanism of DDT's interference with the metabolism of birds and the analytical problems involved in determining the concentrations of DDT in animal tissue. He was on the stand for several days, and the subjects he raised occupied much of the rest of the hearing. The analytical work was crucial, for it established the basis for the environmentalists' claims that DDT residues were found throughout the environment, and the intervenors made strenuous efforts to discredit this research. They tried, by cross-examination and by the testimony of their own witnesses, to show that polychlorinated biphenyls (PCBs) were a major contaminant that interfered with the analysis for DDT. These compounds, similar in structure to the chlorinated hydrocarbon insecticides, were persistent and were found, as was DDT, in analytical studies of ecosystems throughout the world. PCBs, the intervenors claimed, could not be distinguished from DDT. Hence the results showing massive DDT contamination (presumably from bioconcentration) were useless. The subject brought a welter of confused and conflicting testimony, all of it technical and much of it repetitious.

Risebrough's testimony on the formation of eggshells and the biochemical mechanism of DDT's interference with liver function was just as important, for if the petititoners could establish that DDT actually interfered with avian reproduction, it would be grounds for declaring it a pollutant and forbidding its release into the waters of the state or any use that would release it into the ecosystem. Before the EDF could establish the mechanism of DDT's action in birds and mammals, though, it had to show that there was substantial contamination in the state. Risebrough's testimony was interrupted to allow Joseph Hickey to testify about his studies of residues in the environment, particularly the contamination of the Lake Michigan ecosystem with pesticides.[23] Having established that animals were concentrating DDE in food chains in Wisconsin, the EDF recalled Risebrough to show how DDE affected the birds' system. Experiments, he said, had shown that DDE injected into a bird's breast muscle reached the liver and stimulated the production of hepatic enzymes that broke down hormones, including estrogen. (This was Peakall's work.) The estrogen level controlled the storage of calcium in medullary bone, the source of calcium carbonate for eggshells. Risebrough's testimony was as short, his cross-examination as long, and his testimony as crucial as Wurster's. If allowed to stand, it established a mechanism by which DDT could be held responsible for the decline in some bird populations. It would also show that DDT had biochemical effects, casting doubt on the intervenors' contention that there was no danger to man because there were no clinical symptoms of poisoning. As Risebrough pointed out, "there was no mechanism in the establishment of tolerance level to consider this enzyme inducing phenomonon in people."(679)

In an effort to shake this testimony McLean cross-examined Risebrough for most of three days. He began by pointing out that the disappearance or decline of species was not well understood, and he cited an article from the 1888 edition of the Encyclopedia Britannica to show a natural decline in the peregrine falcon population in England in the last half of the nineteenth century. Risebrough said that this article was, as ornithologists had known for years, in error. (Hickey confirmed this point. [1145]) McLean moved on to the question of chemical analysis for DDT residues. Finding that Risebrough's degree was in molecular biology, he asked if Risebrough considered himself a molecular biologist or an analytical chemist. Risebrough replied that he did not "believe in pigeonholing people. I consider myself . . . as an environmental scientist. And I think it's precisely because people have considered themselves specialists that very few people realize what's going on in the environment."(704)

McLean then tried to show that the detoxification mechanisms in the human liver broke down ingested DDT, citing as proof a paper on the occupational exposure of factory workers to DDT. This gave Risebrough the chance to point out the environmentalists' objections to the standards of these studies. It had, he said, omitted any but fully grown males, and the researchers had made no biochemical tests, no steroid tests, no protein-bound iodine tests—the basic test of metabolic functions—no liver function tests, and no blood function analyses. It was impossible to show from this paper that DDT was safe for the general population.(762–789) McLean introduced quotations from a letter by Eugene H. Dustman, director of the Bureau of Sport Fisheries and Wildlife (U.S. Department of the Interior), that seemed to cast doubt on the validity of Risebrough's research methods and the conclusions of his study. The EDF produced a copy of the letter (exhibit 77) from Hickey's files and showed that McLean had quoted Dustman out of context. McLean lamely admitted he had never seen the whole letter. He said that someone—he did not say who—had shown him the letter or a portion of it while he was in the hall outside the hearing room. He had based his examination on this excerpt.(861–865)

At the end of Risebrough's cross-examination, Van Susteren adjourned the hearings for a month, until 13 January. Although the EDF had not completed its case, several points were already clear. One was that the industry could not afford to dismiss the environmentalists as sentimental nature lovers: they had gathered an impressive group of witnesses to prove their contentions that continued use of DDT could have disastrous consequences and that present levels were dangerous to wildlife. McLean's efforts to associate the Environmental Defense Fund with the lunatic fringe had failed, and his cross-examination of the petitioners' scientists had been completely ineffective. It was equally clear that the environmentalists were dominating the public record; the news stories carried, almost every day, some new testimony against DDT. Part of the industry's failure here had to be laid to lack of preparation. Although McLean promised reporters early in December that he would produce witnesses to refute all the testimony the petitioners offered, he could not name them or produce any evidence for his case. The industry did not hand out press releases or brief reporters. Not until January did it make any efforts to get its story across to the public. Much of the EDF's domination of the papers, though, was due to its own attitude and preparation. Its policy of open meetings and briefings gave the reporters something each day, and Yannacone, who was determined to try the case in the papers as well as the hearing room, had his own strategy. Each morning, well before Whitney Gould's 11:30 deadline, he would try to introduce some interesting quotation or bit of information into the record. It would then appear in that day's *Capital-Times* and on the evening news, whereas any rebuttal from later in the day would not come out until the *Wisconsin State Journal* hit the streets the following morning.[24]

Effective influence on public opinion was probably beyond the intervenors' grasp, even though McLean had been a skilled propagandist. Years of argument, counterargument, and propaganda had given the public a great mistrust of DDT. People were ready to ban it, and the signs were apparent even before the hearing began. Early in November the Wisconsin Department of Agriculture and the University of Wisconsin had announced that they would not recommend DDT for the control of Dutch elm disease in 1969. They were defensive about previous recommendations, though the announcement denied that the coming hearing had had any effect on the decision. Guidelines for insect control, the statement said, had never been a "single, simple formula" and each community had been encouraged to develop its own program. "Unfortunately," it went on, "many community control programs have relied too heavily on the use of insecticude sprays. . . ." With the increased public concern about pesticide residues, the availability of other control measures, and "the general lack of appreciation by the public for the continued need of pesticides," to continue recommending DDT for an "aesthetic problem" would "place in jeopardy the availability of this and other essential materials."[25]

There were other indications. The Chilton (Wisconsin) Rotary Club (probably not a radical organization) sent a letter to the Department of Natural Resources a week after the hearing started, asking that the department "outlaw the use of . . . DDT . . . because the use . . . is, without question, polluting the environment of this world and killing its wildlife and . . . is unconscionable and beyond forgiveness."[26] Students from the university also became involved. On 11 December, the Science Student Union distributed a flyer—"The People vs. DDT"—which cited the poisonous effects of the chemical, including possible estrogenic and mutagnic effects. The students proposed to add their voices to those of the scientists "to ensure that the case against DDT is heard over the objections and harassments of the chemical corporation lawyers."[27] The next day the "DDT Commandos" marched from the university to the Capitol, carrying signs saying "Ban the Bug Bomb" and "Liberate the Ecosystem." They marched around the Capitol chanting "We Hate Bugs," spraying people and grounds with squirt guns to "characterize the callousness of the industry which manufactures DDT."[28]

A week before the hearing reopened, the Capital-Times's "Question of the Day" was "Should the Use of DDT be Banned in Wisconsin?" The paper printed four affirmative replies, none from rabid conservationists, all showing a great uneasiness on the subject. On this cheerful note McLean returned to the battle. He had, at least, the legal aid he had asked for; two attorneys appeared with him: Frederick S. Waiss, a San Francisco lawyer employed by Stauffer Chemical Company, and Willard S. Stafford, a Madison trial lawyer. Waiss conducted the examination of industry witnesses for the intervenors, McLean secured witnesses, and Stafford took over daily conduct of the industry's case. Stafford, "who looks and sounds like the guy sent over by central casting to play the lawyer," was a first-rate trial lawyer. He threw himself into the case, but he could do little. It was too late to restrict the scope of the testimony to work done on Wisconsin; the EDF had already made their case on the basis of worldwide contamination. Although Stafford objected and resisted (to the point that the examiner simply cut off the record whenever Yannacone and Stafford began to argue), he was fighting an uphill battle and he knew it. "They would have stoned John the Baptist down on State Street if he came out for DDT," he is reputed to have said. The Industry Task force for DDT had hired him for his courtroom experience and knowledge of local law, not for his scientific experience; he had spent his Christmas vacation studying scientific texts. The industry did not provide scientists.[29]

When the hearing resumed, the EDF turned from birds to fish, calling Kenneth Macek, a fisheries biologist working at the Department of the Interior's Fish-Pesticide Research Laboratory at Columbia, Missouri. Macek, who appeared as an official witness for the Fish and Wildlife Service, testified about experiments on DDT's effect on reproduction in brook trout and on the ability of DDT-fed fish to withstand stress.(971, 1000) Exposure of either parent to DDT increased fry mortality, and fish with DDT in their fat were less resistant to stress. McLean hardly challenged Macek. He had no studies that proved the contrary, or even cast doubt on the testimony. His only point of attack seemed to be Macek's tentative conclusion that indicated a trend of increased growth in DDT-fed male brook trout.(1011–1013) On redirect examination, Macek said that the trend, which he was still investigating, showed that the DDT-fed fish laid down more fat tissue, thus becoming more efficient at accumulating DDT. This, he pointed out, made the fish still more susceptible to environmental stress, particularly to temperature changes.(1012–1023)

The next witness was Richard Welch, a biochemical pharmacologist employed by Burroughs Wellcome Laboratory, who had, as part of his work, investigated the effects of chlordane, DDT, and DDT-derived compounds on the metabolism of steroids in mammals. He had found that the insecticides caused "a marked increase in the rate of metabolism of several physiologically important steroids by liver enzymes," the same effect Peakall had found in birds.(1044) Other researchers, he said, had found similar increases in enzymatic activity caused by DDT in rabbits and pigeons.(1059) Welch's testimony raised again the issue of enzymatic effects as compared to pathological ones. What constituted "proof" of DDT's safety? One exchange between McLean and Welch over two studies on the pathology of DDT illustrated the difference between the two sides.

> *McLean:* I say these studies tend to prove the lack of significance of pesticide residues in the population as a whole, one paper; and the other tends to prove the lack of significance of massive exposure to DDT to men who are engaged in its manufacture.

> *Welch:* No. . . . It proves that at a tissue level of 10 parts per million there are no overt pathological changes in the organs. This is quite different from any enzymatic changes that might be present in these organs.(1116–1117)

The testimony produced some of the most startling head-lines in the hearing, as one of Yannacone's attempts to make the hearing relevant to the public proved exceptionally successful. One of the affected hormones was testosterone—male sex hormone—and Yannacone asked Welch if these hormones were the same in rats as in man. Welch said they were.(1072) When Whitney Gould wrote her story on Welch's testimony, the *Capital-Times* copy desk put on the headline: "Scientist Warns of DDT Peril to Sex Life." The *Wisconsin State Journal* picked up the story, using a similar tag: "Scientist Fears DDT can Cause Sex Change." Even the *New York Times* said: "DDT Termed Peril to the Sex Organs."[30] (Gould, who knew how misleading the headlines were, spent the next week protesting to the scientists that she had not written the headline, only the story, and that was accurate.)

Welch had laid the groundwork for testimony on studies of birds. Hickey returned to testify on the work he, and others, had done to determine the cause of recent population changes in birds. He reviewed his studies on urban songbirds subjected to DDT sprays for Dutch elm disease, and on residues of DDT in soils and in Lake Michigan ecosystems, but his major contribution was a discussion of the data on the decline in certain birds of prey since 1945. He chronicled, with alarming statistics, the decline of various species of hawks and eagles and the complicated scientific work scientists had done that had convinced many that DDE, which they had until then regarded as a harmless, universal contaminant of the ecosystem, was to blame for the thin eggshell syndrome in the raptorial birds. "[I]n my scientific judgement," Hickey said, "DDE is a chemical compound of extinction."(1212)

The next witness, Lucille F. Stickel, pesticide research coordinator at the Patuxent Wildlife Research Center, testified on feeing studies that linked the field observations and the physiological work. Comparing the eggs of ducks fed small amounts of pesticides with eggs from control groups, she said, the researchers had found dramatic differences. Mallards produced 23.5 percent cracked eggs when fed a diet containing 10 to 40 ppm DDE, against only 4 percent for the controls. Uncracked incubated eggs from the DDE-fed birds produced less than half as many healthy ducklings as did eggs from the controls. American sparrowhawks fed 2 ppm DDT plus ⅓ ppm dieldrin or 5 ppm DDT or 1 ppm dieldrin showed 15 percent thinning of the eggshell.(1218–1220) "From these studies," she said, "it is evident that DDE in minute amounts can cause marked impairment of the reproductive success among at least two major bird groups."(1221) Her presentation was short, but very important. Risebrough and Welch had shown that DDT had enzymatic effects in the liver and Hickey had testified to observed correlations between residues and reproductive failure. Stickel's testimony on controlled experiments in the administration of DDE to birds "nailed down" the connection between DDE and thin eggshells,[31] establishing that the relation between DDE and thin eggshells was one of cause and effect.

The appearance of Stickel and Macek, both official witnesses for the Department of the Interior, shows another advantage of the hearing. Neither scientist could, or would, have been a speaker at an environmentalists' rally against DDT, but in Madison they could appear and testify without appearing as advocates. Witnesses were simply impartial experts helping the hearing examiner reach an informed judgment. The neutrality of the hearing guarded them against both Congress and their peers, who generally regarded public advocacy as inconsistent with the pursuit of science and censured those scientists who became involved in such matters. The format of the hearing undoubtedly attracted some scientists, and allowed others when called to testify. It is not, though, the only reason for the EDF's success in finding witnesses. In sharp contrast to the position of Carson's defenders in the immediate aftermath of *Silent Spring*, the EDF in 1968 had little trouble finding scientists. Its problem was avoiding hurt feelings in those who were not needed or who were not put on the stand because they would not make good witnesses. By 1968, the scientific community was coming to support the environmentalists' position, and the EDF had little difficulty in finding scientific help.

Robert Rudd, a zoologist from the University of California, Davis, was the last witness, and a logical choice to finish the EDF's direct presentation. His book, *Pesticides and the Living Landscape*, had summarized the scientific evidence of the effects of pesticide residues on ecosystems, and he provided the broad overview that the EDF wanted to close out its case. It was clear, even before Rudd took the stand, that the environmentalists had prepared by far the best case against the chemical that had been made. The industry clearly would have to mount an extraordinary effort to clear DDT.

CHAPTER 8

Is It Safe and Necessary?

> So far the hearings have been rather dominated by
> a couple of anti-establishment lawyers from New
> York and a beardy from Berkeley.
>
> Robert E. Tracy[1]

WHEN THE EDF finished its case, Van Susteren granted the
intervenors a continuance to allow them to prepare their
rebuttal, but, when the hearing resumed on 29 April, it
quickly became clear that the pesticide manufacturers had not
used the time to prepare a case against the environmentalists'
scientific witnesses. They relied on the same case they had
used in the past, stressing DDT's contributions to public
health and agriculture, the need for the chemical in emergen-
cies, its low cost and great effectiveness, and the absence of
illness or death among people exposed to it. The only new
element was that economic entomologists testifying for the
intervenors tended to downplay the continuing use of DDT
for routine insect control. Some of the testimony indeed gave
the impression that a ban would be unnecessary—that the
chemical was being phased out. DDT's defenders did not
directly address the environmentalists' case; they introduced
no testimony to counter the evidence on the enzymatic
activity of DDT, the population crash of the raptors, or the
links between DDE and thin eggshells.

By April it was clear that this would not be enough.
Pressure was building for a ban, and only a complete vindication of DDT could stop it. The Wisconsin Natural Resources
Board had announced that it "would issue no permits for the
use of DDT during the plant growing season" and would
discourage its use during the dormant season. The board gave

as its reason the "long-range persistence of DDT and its associated harmful effects on fish and game and its pollution of our environment."[2] It was the first official indication that the state agencies recognized the dangers of DDT and would act to halt its use, and it was another blow to the pesticide industry's claims that responsible authorities saw no danger in DDT. In February the industry suffered another setback when two state senators, Fred Risser and Martin Schreiber, introduced a bill banning the use of DDT in Wisconsin, and twenty-seven assemblymen sponsored a similar measure in the lower house. There was little open opposition; opponents backed an alternative, a bill to create a pesticide review board.[3] The hearing was even attracting attention outside the state. The day after the intervenors began their presentation, the *New York Times* noted that "in the last few weeks the anti-DDT battle—which heretofore had been fought with a lot of words and a little action—has taken a new direction that threatens the life of this deadly chemical." After discussing bans in Sweden and Michigan and a suit filed in California, the story went on to say that "currently a hearing on a citizens' petition to outlaw DDT in Wisconsin is serving as a national public forum. Science and industry are presenting arguments that may ultimately decide whether the issue will be settled once and for all with DDT bans instituted across the country."[4]

Other events put more pressure on the intervenors. In March the United States Food and Drug Administration seized 22,000 pounds of coho salmon in Michigan on the grounds that it was contaminated with such high levels of pesticides as to be unfit for human consumption. There was no established tolerance level for residues in fish, but these had up to 19 ppm DDT and dieldrin, almost three times the 7 ppm that was the limit in most food products. The action threatened Michigan's attempts to build a revived fishing industry based on the coho, which it had introduced to replace the lake trout almost wiped out by lampreys. Both sport and commercial fisherman were enthusiastic about the new fish, and officials estimated that in a few years anglers coming into the state to catch coho would spend $100,000,000 a year.[5] Faced with a disaster for one of the state's biggest industries, tourism, the state government reacted quickly and drastically. The Michigan Department of Agriculture quickly cancelled all DDT registrations—the only exceptions were for mice, bats, and human body lice, uses that would require only small amounts of the chemical. It was a significant victory for the environmentalists, even though the men who took the action, including Michigan Secretary of Agriculture B. Dale Ball (one of the defendants in the EDF's early dieldrin case) were not concerned about dead robins or ecological systems.

Michigan was not alone. Across the lake, the Wisconsin Resources Conservation Council distributed a handbill warning that fish from Wisconsin waters might be harmful to human health, an action one legislator condemned as "blackmail," and which Vilas County extension agent Herman Smith claimed was an attempt to kill the recreation industry. The legislature, though, took it seriously and Governor Warren Knowles called for control of hard pesticides to prevent the destruction of the lake.[6] Other states also began to debate the wisdom of a ban on persistent pesticides. More important was the rising public concern about the effects of human action on the environment. By fall, committees across the country would be preparing for the first Earth Day, 22 April 1970, and the environment would be a political issue. The work of public education that the EDF had undertaken was now superfluous; the public was alarmed and ready to act.

The rising groundswell of public support was welcome, but the EDF faced immediate problems that no amount of good will could solve. The $25,000 the CNRA had raised in December was gone, and the organization had to find another $25,000 to keep the action going. The Rachel Carson Fund of the National Audubon Society contributed, but it was no substitute for an assured source of funds. The EDF was still living hand-to-mouth.[7] Yannacone and Wurster were also finding the hearing a serious burden. In April the EDF had to hire another attorney to take care of Yannacone's practice—

his only source of income—and Wurster was having to slight both his students and his own professional work. By summer he warned the other trustees that this could not continue; he needed help.[8]

Little of the strain, though, was apparent in the early spring, as the defenders of DDT called to the stand their first witness, Wayland J. Hayes, professor of toxicology at Vanderbilt University and former chief of the toxicology section of the U. S. Public Health Service (1949–1968). Hayes appeared to testify about the Public Health Service studies done under his direction.[9] His work had been the mainstay of the pesticide industry's contention that DDT was safe, and the intervenors were obviously counting on it, and Hayes's distinguished record, to reassure the public and convince the examiner. After putting Hayes's qualifications and awards in the record, Stafford asked him if DDT was safe. Hayes said it was and introduced his studies to prove it. Tests on convict volunteers who had ingested large doses of DDT daily for periods of up to a year and studies of men who had worked in DDT-manufacturing plants for up to eighteen years showed that these men, despite elevated levels of DDT in their blood and fat, had no signs of DDT poisoning. Stafford inquired about the effects of environmental levels of the chemical. The Food and Drug Administration, he said, had just seized ten tons of coho salmon contaminated with up to 19 ppm DDT. Were these fish a health hazard? Hayes pointed out that one would have to live on them for nineteen years even to get the same dosage that workers in DDT-manufacturing plants received, and they had shown no ill effects. What about possible estrogenic effects of DDT? Hayes discounted this possibility; very high doses produced only temporary effects in rats. (1411–1414) After leading Hayes through a survey of the possible dangers, Stafford again asked if Hayes thought DDT was safe for the general population over a lifetime. Yes, he said, it was. (1422–1423)

Hayes had been an impressive witness—experienced, poised, and prestigious—and his testimony provided the intervenors with a formidable argument against the petitioners and anti-DDT groups throughout the country. Although his studies did not touch the case against DDT made on the basis of its effects on wildlife, they made it more difficult to use that argument. It would be much harder to arouse public opinion against a material that was harmless to man—whatever the effects on wildlife—than against one that might be poisonous to humans. Unless the petitioners could discredit Hayes's testimony, they might win a legal victory in Madison without seriously affecting the rest of the country. Worse, failure to meet this issue would leave them open to the charge that they were more concerned with wildlife than with human lives or food.

The EDF tried to cast doubt on Hayes's studies by stressing the limitations of the medical tests and by citing animal and in vitro studies that suggested that DDT might have harmful effects on man. The tests, Yannacone said, showed only that occupational exposure to or the ingestion of large amounts of DDT would not produce clinical symptoms of poisoning in healthy, adult males. The test groups had included no infants, old or sick people, women or others who might react differently to DDT than did the test subjects. Had Hayes run tests to see if DDT affected the production of hormones, or if it had affected neuro-physiology? Had he tested the relation of dosage to storage, checked the possibility of mutagenic and enzymatic effects? Was he aware that even low levels interfered with the biochemical functions of the body? What about the detoxification of DDT by the liver, particularly in infants? (1460–1508)

Cross-examination clearly pointed up the differences in the way the two sides defined "safety." Hayes's studies, which the intervenors accepted as showing the clinical effects of DDT, had been designed to discover symptoms of pathological changes in bodily organs. They were classical medical studies, organized around the medical concepts of disease and bodily disfunction. "Safety" meant that the material did not produce clinical symptoms of illness. The petitioners approached the subject from an entirely different standpoint. Their concept of "safety" was not tied to pathological changes, but to biochemical and enzymatic effects. They did not use the criteria of clinical medicine, but those of physiology and biochemistry. So long as the biochemical effects of DDT remained unexamined, the safety of DDT could not, they believed, be established.

Despite, or perhaps because of, Yannacone's aggressive pursuit of Hayes, the cross-examination did not produce a wave of headlines that might have shaken public confidence. The Capital-Times said that "DDT Advocate Admits He Didn't Make Key Liver Test," but it was the only newspaper to indicate deficiencies in Hayes's preparation.[10] In general, reporters concentrated on the clash of personalities. If the examination did not influence the public, it did at least shake the intervenors, who regarded it as improper behavior. Hayes himself was disturbed by the rough questioning, and both industry and the EPA were to complain later (at the EPA hearings in Washington) that the rough cross-examination industry and agricultural research witnesses suffered in Madison made it difficult to secure testimony for later hearings.[11]

The USDA also made an attempt to settle the safety issue. Harry Hays, director of pesticide registration in the U. S. Department of Agriculture, entered an appearance "as interests may appear" to put in the record the process of pesticide registration under federal law. Harry Hays, McLean said later, was the "world's worst witness," and his appearance a disaster.[12] Easily led, quickly flustered, and inclined to lose his temper, Hays made a bad impression, and his appearance allowed the EDF to reinforce its own case against current federal regulation. Under cross-examination he admitted that the USDA did not test materials, but simply checked the data submitted by the manufacturers. Nor did it try to insure against all dangers from pesticides. Hays admitted that the department had not required tests of the effects of pesticide sprays on wildlife until 1968. (1535) It did not take into account new scientific studies, and there was no provision for public participation in the regulatory process, no way to initiate procedures for revoking the registration of a pesticide outside the Department of Agriculture. Private citizens did not have access to registration data, nor could they request it. (1551–1558, 1652–1653)[13]

Hays's appearance was a tactical error by the USDA and a blow to the intervenors. It allowed the EDF to spotlight alleged deficiencies in the department's registration of pesticides and, by so doing, undermined the manufacturers' claim that federal regulations proved that materials on the market were safe. The *Milwaukee Journal* headed its story of 1 May, "No Double Check on DDT Data; U.S.," and reproduced verbatim some of the dialogue in which Hays admitted that the Department of Agriculture accepted without tests the manufacturers' data.[14] Before Hays testified, the industry could confidently allude to federal regulation as an effective safeguard for the public; afterward it was more difficult.

The Harry Hays episode is a good example of testimony aimed at the community, not at the examiner. Federal regulation was completely irrelevant to the main issues of the hearings—the biological and chemical effects of DDT and its metabolites on organisms in the ecosystem of Wisconsin. The testimony was simply public relations, some soothing words about the vigilance of the federal government; Yannacone's cross-examination was equally far from the issues. Whether or not the Department of Agriculture was responsive to public pressure or to legal challenges within the federal court system had just as little bearing on the issue. It was important only for the Environmental Defense Fund's case against the current system of regulation.

Wayland Hayes was the only witness offered by the intervenors on the subject of public health, and with his studies in the record, they turned to the uses of DDT, bringing several witnesses to testify that the chemical was needed for pest control. The first was R. Keith Chapman, an entomologist at the University of Wisconsin, Madison, and the author of the critique the entomology department had circulated in the wake of the *Silent Spring* controversy on the campus. Chapman, who had worked with DDT since 1944, testified on its use in Wisconsin and its economic importance to the state. He showed slides of crops damaged by insects, said that chemicals were necessary for insect control in agriculture, talked about the dangers of replacing DDT with other pesticides, and pointed out that DDT use in the state was declining. This was a distinct retreat from the earlier defenses of DDT offered by the manufacturers and by the National Agricultural Chemicals Association. For years they had touted DDT as absolutely essential to agriculture and public health. Chapman, however, left the impression that DDT was being replaced by other materials and would remain, if no action were taken, as a little-used weapon in the entomologists' "arsenal." The intervenors also called Bailey Pepper from Rutgers University, who testified about the range of methods entomologists used to control insects, the extensive trials they had made of alternatives to chemical control, and the failure of these methods. The abolition of DDT, he said, would create a gap in insect control that existing materials could not fill, and he urged that it be retained as a useful reserve.

Makers and users of DDT also came to defend the product. Samuel Rotrosen, president of Montrose Chemical Company of California, one of the last manufacturers of DDT in this country and the main support of the Industry Task Force for DDT, came as an intervenor witness. He suggested that DDT, whatever its status in the past, was becoming less important in chemical control of insects and in the environment. Seventy percent of current production went abroad, and of the 40,000,000 pounds used annually in this country, only 27,000,000 were used for agriculture. Twenty million of these went to cotton production, where the residues, he said, would not reach the consumer. (2468-2471) He returned to the familiar theme that DDT was vital for the control of malaria. The U.S. Public Health Service annually used 2,500,000 pounds of DDT in its *Aedes ageypti* control program, and, together, the Public Health Service and the World Health Organization had used over 75,000,000 pounds of DDT for insect control. (2471–2473) DDT, he suggested, should be retained for its value in public health insect control programs. Taft A. Pierce, an entomologist and vice-president of Orkin Exterminating Company, said that DDT was the only material available for the control of mice in buildings, where it was used as a 50 percent tracking powder, and for bats, where it was sprayed in their nesting places. (2512–2519) The amounts involved were small—Orkin had used only 80 pounds the previous year—but DDT was necessary.

Rotrosen's testimony was window dressing, for he was not technically qualified to speak on the problems before the examiner, and was, he admitted, unfamiliar with the submission of data to the Department of Agriculture for the registration of DDT; he did not even know when the company had last submitted data. He had no more than a layman's knowledge about insect vectors of disease, and, since the intervenors had not shown the presence of any disease in Wisconsin spread by insects, the entire subject was irrelevant. Indeed, there was very little testimony at any of the hearings on the use of DDT to protect human health in the United States, and at the Environmental Protection Agency's hearing in 1971–1972, Surgeon General Jesse Steinfeld said that he knew of no public health use of DDT in this country for twenty years.[15]

This part of the case was at least familiar, but in Madison the intervenors also had to undertake a new task, a scientific attack on the evidence of environmental degradation. It was here that the industry's lack of preparation and adequate scientific advice told most heavily. Problems began with poultry science. On the morning of 5 May, Stafford brought to the stand Frank Cherms, a professor of poultry science at the University of Wisconsin, in an attempt to cast doubt on the work of Hickey and Stickel. Poultry, though, were not closely related to raptors, and the experimental conditions Cherms had used bore little resemblence to those Stickel had set. In addition Cherms admitted under cross-examination that it was "somewhat out of my area within—of competence, I think, to get into the biochemistry and so forth of eggshell formation." (2199) The EDF found Cherms's testimony more useful than did the intervenors. In a brief submitted at the end of the hearing, it stated that "the failure of the agricultural community to recognize the ecological disaster resulting from the widespread use of DDT was exemplified by Dr. Cherms's description of his major research effort," and it went on to quote several pages of testimony.[16]

During Cherms's testimony, the intervenors suffered another interruption, as the public intervenor, Robert McConnell, brought in a witness, Goran Lofroth, a radiobiologist and radiation chemist who headed the working group on environmental toxicology of the Ecological Research Committee, a group set up by the Swedish National Research Council. Lofroth had come at the request of the public intervenor, to testify "as interests may appear" about his committee's review of the data on the toxicity of DDT. His appearance caused a long dispute, for he was not, at least in the intervenors' view, a neutral witness. The EDF had made contact with him in Sweden and had paid his expenses to Madison,[17] although McConnell, who could introduce witnesses in the middle of the intervenors' case (which the EDF could not) had sponsored his appearance. Stafford said that the entire proceeding was improper. Lofroth, he claimed, was the petitioners' witness. Yannacone, however, piously denied the charge. Lofroth, he said, was not there to present a case for the EDF, but merely to introduce data relevant to the hearing. Certainly the EDF was interested in his testimony, but only because it wanted to hear all the facts. (2013–2014) The two attorneys continued this running argument during Yannacone's examination, with Stafford objecting that it was an improper extension of McConnell's direct examination.

Lofroth provided the most dramatic testimony of the hearing. After reviewing data from around the world on the concentration of DDT in human tissues, he talked about DDT in human milk. The average concentration, he said, was 0.1 to 0.2 ppm, and a nursing infant probably drank 15 grams per kilogram of body weight per day. This was a dose of 0.02 mg. of DDT per day, twice the recommended daily intake set by the World Health Organization, and within the range in which laboratory animals showed pharmaco-dynamical changes. (1953–1954) There was evidence that the present exposure level interacted with the body, causing biochemical reactions that interfered with normal processes. There was, he stressed, no way of predicting consequences, and "no significant scientific evidence that DDT compound [sic] is safe for man with the present exposure levels." (1976) Although Stafford tried hard, he could not shake Lofroth's testimony. This was partly due to the impressive research of Lofroth's group, but more because the major thrust of his testimony was not that somehing was wrong but that there had been too few studies to assess the potential for harm. Stafford's examination only strengthened this impression. He asked at one point if Lofroth knew of any incident in which children, breast or bottle-fed, had been harmfully affected by DDT. Lofroth replied that "to my knowledge there has been no investigation of the thing even. That's even worse." (2021)

The press had a field day. The *Milwaukee Sentinel* headed its story of 6 May "DDT in Mother's Milk, Swedish Scientist Says," and the *Wisconsin State Journal* printed "Doctor Tells DDT Danger for Infants." These headlines, just as disturbing as those on Welch's testimony ("Scientist Warns of DDT peril to Sex Life"[18]) and more accurate, were the biggest publicity boost the EDF got during the hearing. They also put the Industry Task Force for DDT back on the defensive and took attention away from their case. The effect was similar to earlier publicity about radioactive strontium-90 in milk. It made the issue an extremely personal one. The EDF later used Lofroth's findings in newspaper ads appealing for support. It was not a scientific appeal, but it worked.[19]

When Stafford resumed his presentation, he moved directly to another attack on the petitioners' evidence that DDT was responsible for major reproductive failures in several bird populations. In examining Francis B. Coon, head of the chemistry department of the Wisconsin Alumni Research Foundation, he tried to show that uncertainties in the analysis of eggs and birds' tissues made it impossible to blame pesticides for the thin-eggshell syndrome. Of all the witnesses, Coon had the best claim to be a neutral scientific witness. He was associated with neither side, had participated in none of the preparations, but had helped all the participants. McConnell had asked Coon for information, and Hickey had relied on him for analytical work in his environmental studies. That he testified for the intervenors was an accident. In late November 1968, Ellsworth H. Fisher, an entomologist who favored the intervenors, had approached Coon and asked if he objected to having his name submitted to Van Susteren and McConnell as an expert witness on analytical chemistry. The intervenors had then called on him to testify about gas chromotography.[20]

Stafford sought to show that the analysis for DDT and its metabolites were not reliable because there was considerable interference from polychlorinated biphenyls, commonly called PCBs.[21] Yannacone had a delicate task in cross-examining Coon. He had to dispel the impression that PCB's might be interfering with the analysis, but he could not afford to damage the credibility of the witness. Coon, though testifying for the intervenors, was much more important to the EDF, for his group had done the analytical work for Hickey's studies on the Lake Michigan ecosystems. Yannacone concentrated on the response of the detector to the components of the sample and elicited an admission that for equal amounts of DDT and PCBs the DDT peak would be so much higher than the PCB peak that it would completely mask the PCBs, which would be only a negligible part of the combined peak. (2081–2091) To establish the amount of interference that the Wisconsin Alumni Research Foundation (WARF) had found in routine examination, he asked if the laboratory had found it necessary to prepare a standard disclaimer about the accuracy of their analyses for DDT. Did they routinely notify customers that there was uncertainty in the results of their analysis caused by PCBs? Had Coon notified Hickey or Wurster of any problems in their samples in regard to DDE? Coon said no; the amount and frequency of the interference had not warranted this.(2126–2127)

The results of this episode show one of the problems involved in using scientific testimony in an adversary proceeding. The problems of the analysis were not simple—

indeed, they were quite complex—and each side found points that, taken by themselves, proved its case. Perhaps the confusion and complication are inevitable, but the testimony certainly did not seem clear to the public, and the examiner had to undertake some extra study on the subject. Nor did this, or later, testimony settle the point. Although all witnesses on the subject agreed that there were ways to eliminate confusion in the analysis, the issue of contaminated samples and misidentified materials returned in later arguments and obscured as much as clarified the issues involved.

The intervenors' other attempts to break down the analytical evidence met with even less success. They called Paul Edward Porter, a physical chemist with Shell Chemical Company (a division of Shell Oil), who discussed the breakdown of DDT in the environment and in animals. Porter's main point was the diversity of chemical reactions DDT underwent and the varying speed of the reactions, depending on environmental conditions. Waiss, who examined Porter, obviously hoped to establish the direct and critical interference by PCBs that Stafford had been unable to elicit from Coon. The effort failed, partly due to the examiner's questioning, which brought out ways to eliminate interference, partly due to Porter's willingness to acknowledge the work of others. WARF, he said, was a very good laboratory, "We have the very highest regard for their capability in analyzing."(2224) Waiss then asked about Risebrough. Was the witness familiar with his reputation? "Yes, I am," Porter replied. "He has a very good reputation."(2226) When Waiss finished, Yanacone did not cross-examine. He simply thanked Porter on behalf of the petitioners for appearing to testify.(2240)

After Porter's appearance, the intervenors raised the question of DDT's effects on wildlife population, calling William Gusey, chief wildlife specialist for the agricultural chemicals division of the Shell Oil Company and formerly chief of pesticide surveillance and monitoring, Bureau of Sport Fisheries and Wildlife, Department of the Interior. Gusey, though, had done no studies of the factors affecting reproduction in game animals; his testimony was limited to trends in game populations and a discussion of factors, such as the amount of breeding area, the hunting pressure, and endemic diseases that affected waterfowl.(2252–2283) None of the petitioners' witnesses, however, had testified that DDT adversely affected mammal populations or denied that other factors than DDT might be responsible for fluctuations in wildlife populations. Nor was Gusey qualified to speak on some of the subjects the intervenors raised. At one point Van Susteren cut short the argument between Yannacone and Waiss:

> *Examiner Van Susteren:* All right. Well then, the examiner will ask him. Mr. Gusey, are the papers that your lawyer, Mr. Waiss, asked you about in the same line of work that you are engaged in?

> *Witness:* No.

> *Examiner Van Susteren:* Do you in any way hold yourself out here today as a scientist in regards to specific research in any particular field?

> *Witness:* No, I have never been in research. (2296–2297)

Gusey's appearance was more for public relations than to refute the petitioners, and in this sense it was a success. He did not take the position that industry should be concerned solely with the benefits of its products, an attitude that had characterized much of the early defense of pesticides, and during cross-examination he made a plea for more cooperation between industry and the conservationists. He said that he hoped his new job with Shell Chemical Company meant more awareness of the problems posed by chemicals in the environment. Yannacone concluded his short cross-examination by congratulating Gusey on his attitude and wishing him "lots of luck" in his new job.(2214–2215)

When Stafford closed the intervenors' case on 14 May, it was clear that the industry had failed to stop the rising tide of anti-DDT sentiment in the state or to make a convincing case to the examiner. Caught off guard, lacking scientific advice, aware only too late of the importance of the hearing, it had lost the chance to stop the opposition in Madison. The rest of the hearing would belong to the EDF, and it appeared certain that they would get a local ban on DDT, either through the Department of Natural Resources's decision or through the courts. Never again, though, would the NACA or the USDA underestimate their opponents so badly.

With the conclusion of the intervenors' argument, there remained only the petitioners' redirect case. The environmentalists had already established most of the points they had planned to make—the existence of the ecosystem, DDT's properties in the environment, and its effects on nontarget species—and they devoted the second presentation to countering the arguments that DDT was safe and necessary. On the contrary, they argued, its safety to man had never been established, there was evidence that it was dangerous to mammalian systems, and it could be replaced with safer methods of insect control. To establish these points, they offered more evidence on the effects of DDT on the nervous system, testimony on the general criteria for safety of drugs, and they brought witnesses who were both practical entomologists and advocates of an end to DDT use.

The first witness, on 19 May, was George J. Wallace, who testified about the effects of DDT sprays on robins. He discussed the population studies he had done on these birds, the data he had gathered on bird mortality throughout Michigan, and the analytical work he and his students had done on residues. Their most important finding was the correlation between symptoms of poisoning and high residue levels. All the dead and dying birds had high levels of DDT, and all but one of the robins collected in the sprayed areas had accumulated DDT, whereas robins collected outside the area had none.[22] Statistical analysis of the data, Wallace said, showed a high positive correlation between DDT in the brain and death. Stafford did not directly attack either Wallace's analytical results or his observations; he turned instead to the question of whether the continental robin population was decreasing, citing the work of Phillip Marvin, a consulting entomologist, who based his estimates of an increasing population on the Audubon Society's Christmas bird counts. This argument was an old one, going back to the first defenses of DDT after the publication of *Silent Spring*. In 1963 Marvin had used the Christmas counts to show that, contrary to Carson's warnings, there was no danger that robins would be wiped out by Dutch elm disease sprays. Robert White-Stevens, one of the strongest public defenders of DDT, had picked up the argument over strong objections by Audubon scientists, who did not believe that the counts were any indication of the continental robin population, and said so, repeatedly. Roland Clement had criticized White-Stevens's use of the data and even pointed out the problems to him at a meeting of Sigma Xi in Poughkeepsie. Despite the critique, the Christmas bird counts continued to crop up throughout the discussion of DDT.[23] Neither side offered new positions in Madison.

The question at issue was the neurological damage caused by DDT, and Wallace established a case from field observations. The next witness, Allan B. Steinbach, a research associate in biophysics and physiology at Albert Einstein College of Medicine, continued this line of attack. He confirmed Wallace's observations about the role of DDT in tremoring robins; their behavior, he said, could be accounted for "simply on the basis of the known mechanism of action of DDT on the nerve."(2595) He did more, though, than corroborate Wallace's field observations; he described DDT's effects on the nervous system and his own experiments on the effects of DDT on nerve conductance. After summarizing the history of the study of nerve conduction and nervous system response to various agents, he gave an account of his own work. He had exposed the nerve ganglion of a frog to a weak solution of DDT, and had measured the conductance of the nerve cell directly. DDT, he testified, interfered with the usual sharp potential changes associated with muscle firing impulses in the nerves. Like a telephone system with leaky cables, the nerves sent random or repetitive signals, which

caused muscle tremoring, convulsions, or death.[24] There
was, Steinbach said, no known threshold level, a concentration below which there was no effect, nor were the effects
reversible within the time period of the experiment (six
hours).

This testimony brought back another familiar theme in the
debate over pesticide residues—the effect, or lack of effect, of
small doses. The defenders of DDT were willing to stipulate
that the chemical, in large doses, was a poison, but they
argued that in small doses it had no effect on the body. There
was, they said, a dose-response curve, a correlation between
the severity of the symptoms and the concentration in the
organism. Below some low concentration, the threshold
dose, there would be no symptoms. Steinbach's work
suggested that this model was faulty, that the lack of visible
pathological symptoms did not mean a lack of effect on the
organism. The difference on this question, somewhat muted
during this bearing, was to assume more importance later,
when animal tests suggested that DDT might be a human
carcinogen.

With Steinbach's studies in the record, the EDF recalled
Wayland Hayes for more cross-examination about the safety
of DDT and, to emphasize this point called Dr. Theodore
Goodfriend, assistant professor of internal medicine and
pharmacology at the University of Wisconsin School of
Medicine, to the stand. Goodfriend had done no work on
DDT or its metabolites; he came only to give an expert
opinion on the methods used in the studies and the adequacy
of the evidence on the safety of DDT to man. On the basis of
his knowledge, he said, he could not agree with Wayland
Hayes that DDT was "absolutely safe." He based this on
"instances of blood dyscrasia in humans exposed to DDT . . .
knowledge of neurotoxicity in persons exposed to high doses;
. . . the variability of human response to virtually all agents . . .
the changes in drug metabolizing systems in animals; the
published autopsy correlations of levels of DDT with known
diseases."(2678) Apparent safety, he warned, was not
enough. "[T]he history of endocrinology and pharmacology is replete with instances of compounds that were intro-

duced . . . after what were considered to be adequate tests, which later proved to be toxic . . ."(2691) He admitted under cross-examination, though, that he did not consider aspirin "absolutely safe," which somewhat diminished the impact of his warnings.

Goodfriend's appearance illustrated another advantage of the hearing room as a forum for the environmental case: the EDF was able not only to get scientists who would not care to be identified with a public crusade against DDT, but to get expert witnesses who were not involved with the issue at all. The EDF had needed expert testimony about the physiological tests and the safety of drugs. One scientist knew of Goodfriend's work, Wurster got in touch with Goodfriend, and he agreed to testify. Yannacone asked questions about drugs, testing, and safety; Goodfriend gave the (to a physiologist) obvious answers, and the EDF had another piece of its case in the record.[25]

The EDF wanted to impress on the public that alternative means of control, ones that did not rely on persistent chemicals, were available, that ending DDT use would not mean mass starvation and disease. There were, it argued, methods that would preserve both the environment and man's food supply. To do this it brought to the stand two practicing entomologists who testified that DDT could and should be replaced by integrated systems of control. The first was Paul De Bach from the Department of Biological Control of the University of California at Riverside and the California Agriculture Experiment Station. De Bach had been a specialist in biological and integrated control of insect pests for almost thirty years and had published over 140 scientific papers.(2695–2696) He testified about problems caused by DDT's wide and unselective action. It was, he said, so harmful to so many varieties of insects that he used it to disrupt agroecosystems in order to gauge the effectiveness of natural controls.

Stafford concentrated on possible deficiencies in biological control. He stressed the speed of pest infestations, the alleged slow growth of the predator population, the fluctuations in a pest population even after natural control was established. De

Bach strongly defended his position. Natural controls, he
said, were stable and effective. Infestations varied in speed,
but a species under natural control did not suddenly increase
in numbers. What fluctuations there were occurred at a low
level, well below the economic threshold, the density that
would adversely affect the crop yield. He denied that a predator species only slowly brought an infestation under
control; when a predator was introduced into an infested area,
it normally underwent a rapid population growth.(2721) Van
Susteren also raised an objection, asking if the parasite, once it
had reduced the pest population, would become a pest itself.
De Bach said this was not possible. Insect predators had very
specialized diets, and would not suddenly move into new
ecological niches.(2818)

Donald Alfred Chant, chairman of the Department of
Zoology, University of Toronto, was the next witness. His
testimony was similar to De Bach's, and Yannacone could
have established the effectiveness of biological and integrated
control systems without calling him. It was important,
though, to show that the petitioners had not only scientists,
but respected scientists, on their side. Chant was the Canadian
representative to the Food and Agricultural Organization of
the United Nations and the International Congress of Acarol-
ogy (the study of mites), a member of the Pollution Panel of
President Johnson's Science Advisory Committee (1956–
1966) and of the Subcommittee on Insect Pest Control of the
National Academy of Sciences (since 1965), founder and
continuing member of the National Committee on Pesticide
Use in Agriculture in Canada and a member of the Agriculture
Research Study Group of Science Council of Canada.(2723–
2724)

The EDF ended the hearing with some unusual testimony.
It recalled Orie Loucks, a botanist who had testified at the
beginning of the hearing about the Wisconsin regional ecosystem, to talk about his work in systems analysis. No one had
planned to present systems analysis of the environment when
the hearing opened. Only as the hearing stretched beyond the
original ten days and attracted national attention did the EDF
realize that the hearing offered a better opportunity than it had

anticipated to present a complete record. Loucks formed his
working group in January, with a tentative commitment to
have the group's findings ready for the rebuttal case. By May
the evidence and the model were complete enough to present
at the hearing, and, with the examiner's encouragement, the
environmentalists decided to introduce the evidence.[26] Although there was far too little specific data for a detailed
picture of the environment, and the model was not very
refined, it was adequate for a first approximation. Loucks
found the conclusions quite sobering. "There is no evidence
. . ." he said, "that the breakdown of DDT is equal . . . to the
additional introduction of DDT in the system. It is most
probable . . . that there will be substantially more degradation
of the stability characteristics of the ecosystem if the input of
DDT is not stopped immediately."(2791) The damage did
not depend on the extinction of species already directly
affected by the bioconcentration of DDT. Even small changes
in the predator population, which acted as a check on the
lower trophic levels, could cause large and disastrous changes
in the ecosystem.

   With this ambitious, and somewhat confusing, testimony
(someone was heard to remark that he had never heard of
systems analysis before and hoped he never did again) the
hearing ended. After almost a month of testimony spread
over half a year and the production of a transcript of 2,500
pages and 100 exhibits, it was all over. The hearing examiner
returned to simpler, shorter cases. Lorrie Otto, who had sat
each day in the hearing room, went home to Milwaukee.
Reeder and the others sat down to settle the bookkeeping and
tie up the loose ends. Whitney Gould went off to other
stories. For Yannacone, Wurster, and the other members of
the EDF, though, the end of the hearing was less an end than a
beginning. They had shown that legal and scientific expertise
could be used in a quasi-judicial forum to educate the public,
present their case, attack their opponents, and (they hoped)
end DDT use in Wisconsin. Where were they to go from
here?

CHAPTER 9

Final Rounds

By the time the Madison hearing ended, EDF's situation had changed. When it had begun the action in the fall of 1968, the public had not seemed interested, the government appeared completely unresponsive to the environmentalists' arguments, and funds and support were scarce. The EDF emerged from the Madison hearing into a new world. Foundations were interested in funding its work in environmental law; the Rachel Carson Fund of the National Audubon Society was providing steady support; environmentalists around the country were appealing to it for aid and were, in some cases, raising money for legal expenses. The former "Fundless Environmental Defenders," who had been desperately seeking a case in the fall, found themselves overwhelmed with cases in the spring.[1] The problem now was to choose the most important and useful ones, to make sure that the target was worthy of the EDF's time. Even with regard to DDT, the remainder of the fight, though long and expensive, would not be uphill; the public would be involved, and on the environmentalists' side.

Some of the EDF's support was due to its success, but it also benefitted from the rapidly increasing environmental consciousness of the country. What had been a slowly rising tide had become a tidal wave; 1969 was the year of the environment. Everyone, it seemed, was alarmed at the increasing environmental degradation surrounding him, everyone wanted to do something about it. The environment became an issue second only to the war. Incidents—the Santa Barbara oil spill and the publication of Paul Erlich's *Population Bomb*, for instance—helped, but people were simply ready to believe that the world around them was in trouble.[2] *Silent Spring* had prepared many Americans to accept the idea that human action could be a significant factor in the global ecosystem, to believe that technology could be harmful, and to distrust the experts. By 1969, the accumulation of incidents had turned suspicion to active discontent.

Whatever the movement owed to DDT, by 1969 it had passed beyond exclusive concern with the chemical. Persistent pesticides were a problem and an example of the evil to be stopped, but the movement itself took in a multitude of groups and interests only loosely united around the single issue—the environment—that seemed to cover them all. About the only thing they had in common was the conviction that current policies and trends threatened the further existence of human life, or of any life worth living. Some were primarily concerned about the dangers to the ecosystem from human activity, others about the immediate harmful effects on man. Many of the organizations concentrated on the specific aspect of the problem—overpopulation, wilderness, wild animals, whales—in which they were particularly interested or saw as the key. Some groups saw the environmental crisis as part of their problem. The consumer movement, for example, was one of the original allies of the environmentalists; Consumers' Union printed special editions of *Silent Spring* and, in 1971, of Joseph Sax's *Defending the Environment*. The scientists' movement, which aimed at getting scientists to participate in public affairs was at once stimulated by the environmental movement and was part of the backbone. Public health workers, moving beyond problems of acute toxicity and chronic poisoning, allied themselves with the environmentalists and, in the Reserve Mining Case, involving the dumping of contaminated tailings into Lake Superior, the groups formed a close common bond.[3]

The environmental movement of the late 1960s, though, was much more than a collection of lobby groups; it would otherwise never have managed the political action that it did. The mass movement, which provided the bodies for demonstrations, the volunteers for community action, and the votes for environmental causes, was an outgrowth of the participatory radical politics of the period. People who wanted to form a new participatory democracy, who were anxious to break the power of corporate America, who were searching for a new political focus, or who wanted to adopt a new culture at odds with the American mainstream found in the environmental movement a vehicle for their dreams. Here was an indictment of the practices of America, done by the scientists who had built it; here was the call to a new culture, the promise of living in harmony with nature (and a revival of the cult of the Noble Savage).[4] The movements for small-scale technology, for decentralization of political and economic power, for communal living on the land, for a new nonmaterialist culture—all could embrace the environmental cause, or what they conceived the cause to be. As a result there was no clear ideological, social, or political ground on which all the participants could stand. The adherents and supporters ranged from staid, respectable scientists like Joseph Hickey and solid, conservative organizations like the Audubon Society through the EDF to students deep in the counterculture and engaged in radical politics. Some of the people were committed because of personal experiences or scientific study and worried about the future of the wild areas and things that had played an important part in their lives. Others were "into" the environment as a cause, reflexively against what seemed another manifestation of the sickness of "Amerikan" society. There was also a baffled, sometimes angry group of middle-class citizens, worried about the problem but unsure what could or should be done.

The main impetus for the visible environmental move-ment, the one in the streets, was the students. Accustomed to direct action and demonstrations (usually from work against the war) and committed to changing the "system," they provided strength for the cause. By the summer of 1969, while the EDF was weighing its next moves against DDT, other people were seeking to mobilize the students for action. Senator Nelson of Wisconsin and Representative McClosky of California popularized the concept of a national teach-in, to be held in the spring of 1970, and the idea ballooned into the first Earth Day, 22 April 1970. Throughout the country, millions of college and high school students participated in teach-ins, happenings, and demonstrations, and thousands gathered in Washington to hear speakers proclaim the beginning of a new environmental consciousness.[5]

The public environmental campaign was of no direct help to the Environmental Defense Fund, for its weapons—scientific testimony—and its battleground—the laws and regulations of the states and federal government—were outside the effective reach of demonstrations and teach-ins. It was, though, an important indirect aid, focusing political attention on the problems of the environment, stimulating legislative action, and making government at all levels more receptive to the environmental case and to environmental degradation. A series of actions and incidents stimulated by this public pressure helped change the ground rules under which the EDF would fight the last battles against DDT. In September 1968, the General Accounting Office (GAO) had presented a report to Congress on deficiencies in the USDA's administration of the Federal Insecticide, Fungicide, and Rodenticide Act (FIFRA); it followed that with a supplementary report in February on lindane vaporizers. In April, Secretary of HEW Robert Finch appointed a special commission (the Mrak Commission) to investigate the risks and benefits of pesticide use, and that report became available in November. Congress began its own survey of the USDA's enforcement of pesticide regulations in June, investigating the GAO's accusations, and the rest of the DDT battle would be fought under the constant Congressional threat of amendments or revision of FIFRA. Finally, in late 1969, Congress passed the National Environmental Policy Act, declaring that preservation of the environment was of paramount importance and directing all government agencies to weigh environmental values in considering major projects.[6]

The GAO reports were a severe blow to the credibility of the USDA's Pesticides Registration Division. An investigation showed that, far from vigorously enforcing the law, the division hardly acted at all, and what action it took was probably ineffective. In 1966, for example, it had tested and reviewed 2,751 samples and found 750 in violation—562 of them major violations. The agency, however, made no recommendations for prosecution under FIFRA, and the USDA's recall actions might not, the GAO concluded, have removed all of the offending products from the market. Memoranda from the USDA itself indicated that thallium-containing products (used against commensal rodents) might still be available to the public a year and a half after they had been banned. A separate study of the regulation of lindane, a pesticide used to control insects in buildings, revealed other deficiencies. Despite protests from the Public Health Service, the Food and Drug Administration, and other health agencies, the USDA had registered lindane for use in continuously vaporizing pellets in restaurants and food processing plants and continued that registration for sixteen years.[7]

The reports stimulated a House committee investigation in June and a report followed shortly thereafter. The report confirmed the GAO's findings. The USDA had simply not enforced FIFRA until mid-1967, when a new director of enforcement was appointed. Even then the staff appointed was so small it could not even deal with the backlog of cases, much less pursue new ones. Nor had the agency used the other means of enforcement available to it. The power of suspending a registration, designed to remove a product from the market pending an investigation of its safety, had been used only once, to remove a pest-control strip identical to one manufactured by Shell Chemical Company—hardly a public health emergency. Nor had the USDA worked closely with other agencies to control the health hazards of pesticides. In the wake of *Silent Spring*, the USDA, Department of the Interior, and HEW had signed an agreement, setting up a referral procedure to settle questions about the effectiveness and safety of pesticides. HEW memoranda, the committee found, indicated that that agency had come to believe within a year that the USDA had no intention of living up to the contract. It seems to have been right. All 1,633 of its objections to registrations were filed and never referred to the

Secretary of Agriculture for discussion with the Secretary of HEW (the approved procedure). The USDA claimed referral was not needed, as the situation was resolved by registering the product in question. The USDA had also ignored, for fifteen years, the FDA's complaints about the registration of lindane, claiming that there was no evidence to show that the vapors were harmful, and it waited two years after receiving the scientific evidence to act. It canceled the registration two months after the GAO's report appeared. Nor, despite Harry Hays's claims, did the USDA inform itself about accident statistics—although Hays claimed those were an important tool in enforcement. He had estimated there were about 170 poisonings a year; the committee found that the Public Health Service had files showing about 5,000 a year, which it estimated were less than 10 percent of the cases. There were various other lapses, including conflicts of interest among the consultants and failure to make sure products were removed from the market and that label directions were adequate.[8] The investigations discredited the USDA's role of pesticide regulation, paving the way for a transfer of power to the Environmental Protection Agency, when that agency was formed two years later.

Reports and even legislative action against DDT, though, were dwarfed by a more immediate and alarming issue—its possible carcinogenicity. Suspicions had first been raised in testimony before the Delaney hearings, and again by Carson. The environmentalists had done little with that evidence in Madison, partly because it was not as good as it might have been and partly because the evidence of environmental degradation was so clear. In June 1969, though, the National Cancer Institute finished animal tests it had begun in the wake of Silent Spring. They showed that mice exposed to low levels of DDT for long periods of time had an increased incidence of liver tumors, and though the tests did not prove that DDT was causing cancer in humans, they did raise a serious suspicion and did alarm the public.[9] The FDA had just banned the use of cyclamates in soft drinks, and the Washington Post, in an editorial, "Banning Cancer-Causing Chemicals," discussed the similarity. If the FDA was willing to apply the same rule it had used against the artificial sweeteners—guilty until proven innocent—then "a ban on the use of DDT in its present form may well be in the making." A week later seventeen Congressmen petitioned President Nixon for a ban on DDT, contending that it caused cancer.[10]

Cancer was far more alarming to most Americans than the fate of the peregrine falcon, and the National Cancer Institute's verdict was confirmed by the (HEW) Secretary's Select Commission on Pesticides and their Relation to Environmental Health (the Mrak Commission), which rendered its report in November, just as the EDF began a new battle. The commission concluded that the National Cancer Institute study showed that DDT caused cancer in laboratory animals under the experimental conditions. It did not go so far as to say that this proved that DDT caused cancer in man at environmental levels—indeed, it explicitly disavowed that idea. There was, though, "a remarkable degree of concurrence . . . between chemical carcinogenesis in animals and . . . in man."[11] The commission went on to deal with the ramifications of the problem: the dose-response relationship, the "no-effect" dose (whether or not it existed for carcinogens), the available studies of human populations (the sample size and time were not adequate and follow-up too limited).[12]

The Mrak Commission report should have formed a centerpiece for the subsequent discussion of the carcinogenicity of DDT. It was detailed, carefully done, and comprehensive. Unfortunately, it served the purposes of neither side in the dispute. Environmentalists were convinced (or many of them were) that DDT was carcinogenic and rejected the Mrak report as too limited and conservative. On the other hand, it went too far for the defenders of DDT, who wished to discount the animal evidence (as testimony at the EPA hearing on DDT showed). As a result it was largely ignored, particularly at the EPA hearing, where both sides relied exclusively on the testimony of the scientists who had done the work or were actively involved in the field.

Between the end of the Madison hearing and the early fall

of 1969, while the world changed around it, the EDF underwent a transformation. Reorganization was necessary; it could not continue as a hip-pocket organization, the private preserve of a lawyer and a few scientists. Even the Madison hearing had imposed severe strains on the informal organization and volunteer efforts of the group, with Yannacone and Wurster sacrificing more time and energy than they could afford.[13] The informal democracy of the group had also suffered, as circumstances forced the other members to yield effective power to Yannacone and Wurster, the people on the spot. In June the members approved a set of bylaws that gave the board of trustees (still the only members) the power to set policy and hire a staff to make day-to-day decisions. The board quickly hired an executive director, Rod Cameron. Then, in late summer, the coalition broke up. Under the new bylaws, the original members, who had had indefinite terms, were to elect a new board, presumably themselves, for stated terms of office. At the organizing meeting in late August, the others organized the board without the Yannacones and took in new members from outside Long Island—Roland Clement of the Audubon Society (who had been offered a place on the original board in 1967) and three conservationists. Yannacone was reduced to legal counsel, a hired hand. A month later, after a tense set of negotiations, exchanges of memoranda, and phone calls, the EDF terminated all its connections with Yannacone and hired a new attorney to take over the pending cases.[14]

The rift between Yannacone and the other members was the result of differences, personal and professional, that went back to the early days of the EDF. Common aims, the excitement of action, and Yannacone's undeniable brilliance in the courtroom had smothered the differences for a while, but not permanently. Yannacone and the scientists had different ideas about the EDF. Yannacone saw it in terms of environmental law, of establishing precedents in the field; the others saw it in terms of stopping immediate dangers to the environment. On a more mundane level there were strong differences on working hours and methods. Yannacone carried on an active law practice, fitting the EDF into his spare hours, and his energy, activities, and preferences dictated last-minute, all-night sessions. But what to Yannacone was a necessity—even a pleasure—was, for the others, irritating evidence of slap-dash, careless work, and of Yannacone's lack of consideration.[15] At bottom was a struggle for power (though that is not the way any of the participants would like the matter to be expressed). By the summer of 1969 either Yannacone ran the EDF or the others ran it—along different lines. As a result they forced Yannacone out and fired him.

Bylaws, a director, and a new lawyer did not address EDF's tactical problem—what should it undertake next? Wurster wanted to concentrate on DDT, and during the summer he sent a barrage of memoranda to the other members. The Wisconsin hearing, he said, had been "a splendid success in every way, and this is true no matter how the decision comes out." Now it was time for another suit. DDT was the "world's worst pollution problem"; it was the best documented case the EDF had; there was political support for more action; people understood the issues and were concerned about it. Action was also, he argued, the key to building the organization. DDT had brought in the only money the group had ever raised; it would bring in scientific support as well as more money; decisions against it could be used against other persistent pesticides. In short, DDT was the best issue the EDF had and it was high time to take advantage of it.[16]

There was much to support Wurster's idea. In January, Arizona had instituted a one-year ban on DDT and DDD for agriculture; in April, Sweden had banned the chemical for two years and Michigan had instituted the first full, permanent ban in the United States. In June, Elvis J. Stahr, president of the National Audubon Society, called a press conference to announce the society's support for a ban, and several states were either considering bans or had already taken action on persistent pesticides. Even the USDA was beginning to respond to the increasing public pressure; in August, it severely restricted the use of DDT in cooperative state-federal control programs.[17] The EDF, though, had already taken on other commitments. In the fall of 1968, just as the Wisconsin hearing was coming up, it had discussed the possibility of an air pollution suit against the Horner-Waldorf Corporation's plant in Missoula, Montana with a group called Environmental Defenders of Western Montana. By the summer of 1969, that action was underway, due to come to trial in the fall. In the spring, EDF had taken a case in Florida, an attempt to stop construction of the Cross-Florida Barge Canal. Even as Wurster was urging another DDT suit, the Floridians were raising money for the canal fight.[18]

The expansion of the EDF and its new financial resources—partly its own, partly those of its allies—made it possible to pursue several cases at once, and in the fall the EDF began two new actions against DDT, both aimed directly at the federal bureaucracy and both stressing the dangers DDT posed to man, not to the environment. With the Sierra Club, the West Michigan Environmental Action Council, and the Audubon Society, it petitioned the Secretaries of Agriculture and of Health, Education, and Welfare (the officers responsible for enforcing the 1947 Federal Insecticide, Fungicide, and Rodenticide Act) (FIFRA), asking them to suspend registration of DDT (which would put an immediate end to its use), start cancellation proceedings (which would result in a permanent ban), and lower immediately the tolerance level on DDT in human foods to zero. With the support of California Rural Legal Assistance, it filed another petition with HEW on behalf of five pregnant or nursing women, seeking a zero tolerance level on DDT in food. The chemical, it said, was a carcinogen, and under the Delaney clause the Secretary was required to set a level of zero for such materials.[19]

The petitions marked a shift in EDF's tactics, as the reliance on the cancer issue indicated a change in the argument against DDT. One of the EDF's first suits, quickly abandoned, had been a direct legal challenge to the USDA.[20] The Secretary of Agriculture had claimed that under the federal law, private parties had no interest with regard to FIFRA, hence no standing, and the court agreed. Yannacone had vowed not to try that route again; he sought ways to challenge the agencies without directly confronting the issue of their legal authority and by remaining outside the administrative structure of the federal statutes, to make them fight on EDF's own ground. The cases in Wisconsin and Michigan on dieldrin and DDT had been the result, and the organization had studiously avoided federal suits based on federal regulations. Now EDF was seeking to force the agencies to act under the prevailing system of regulation. The tactical shift was not due just to the replacement of Yannacone with William Butler; it was more directly linked to the new evidence on cancer, which convinced many of the environmentalists that they had to seek immediate action. The EDF issued a press release to accompany its legal action, summarizing the National Cancer Institute findings. DDT, it concluded "was clearly carcinogenic . . . because it caused cancers of the same kind and at approximately the same frequency as did known cancer-causing agents."[21]

The petition to the Departments of Agriculture and to Health, Education, and Welfare began the process that culminated in the Environmental Protection Agency's hearing and in the final ban in June 1972. Progress, though, was neither automatic nor easy; the EDF had to plod through a series of legal and bureaucratic obstacles. In response to the original petition, HEW Secretary Robert H. Finch announced, on 12 November, that the administration planned to phase out all but "essential uses" of DDT in the next two years—by the end of 1971. Concern about the chemical had, he said, grown since the publication of the National Cancer Institute's study earlier in the year, but he stressed that there was no cause for alarm. The USDA did not even go this far. Secretary Hardin announced a ban, to begin within thirty days, on the use of DDT in residential areas and said that the department would issue a notice of intent to cancel these uses in ninety days.[22]

The environmentalists wanted far more. Hardin had only stopped a visible but insignificant use of DDT. Finch had taken no immediate action, and his promise that all but "essential uses" would end in two years was no guarantee that the situation would change—what were "essential uses"? HEW had also rejected the petition from California, and that meant there would be no action on DDT in food. They condemned the government's action as "inadequate, unenforceable, and not commensurate with the nature of the threat," and late in December they filed petitions for review with the U. S. Court of Appeals in the District of Columbia, asking that the court direct the Secretaries to take "swifter and more effective action."[23]

While these skirmishes were going on, the EDF took another step in its transformation from a small group of friends to a national environmental lobby. In March it invited the public to join. The campaign was successful, in part because the organization made a full-scale attack on the public's emotions. One of the first appeals for members was an advertisement covering two-thirds of a page in the *New York Times*, headed: "Is Mother's Milk Fit for Human Consumption?" No one knew, it said, but if the milk appeared on the market, the FDA would confiscate it as contaminated. The EDF, it continued, was a nation-wide coalition of scientists, lawyers, and citizens dedicated to the protection of environmental quality through legal action and public education; it was now, for the first time, soliciting public memberships.[24]

The drive for members was more successful than the appeal to the courts, at least in terms of initial response. A series of court challenges, government answers, and continued legal appeals occupied the rest of the year. Despite setbacks, though, the EDF made progress. In 1968, when the Secretary of Agriculture had claimed that outside organizations had no standing to challenge the Secretary's rulings under FIFRA, he had found a sympathetic audience on the bench. In 1970 he did not. In May, a three-judge panel of the U. S. Court of Appeals in the District of Columbia disagreed with Secretary Hardin. Judge Bazelon, speaking for the panel, said that FIFRA allowed standing to anyone affected under the act, not merely to the formulators and manufacturers of agricultural chemicals. He ordered the USDA to suspend all DDT use within thirty days or justify its failure to do so. That same day another panel of the court ordered HEW Secretary Finch to publish in the *Federal Register* the environmentalists' proposal for a "zero tolerance" on DDT in raw food.[25] A few weeks later, the EDF got its first support from the administration, as Secretary of the Interior Hickel banned many pesticides from federal land. The widespread flouting of current regulations, he said, showed that DDT was not being safely used. The USDA, however, remained intransigent. In response to the court of appeals, Hardin reaffirmed his decision not to cancel or suspend registrations. DDT, he said, was not an imminent hazard to human health or to fish and wildlife, and it was essential for some agricultural uses. A few months later, the USDA did announce the beginning of cancellation procedures for over fifty uses of DDT, but it did not include cotton, which accounted for over 75 percent of the DDT used in agriculture. In January, EDF had warned its friends that "a long battle may yet lie ahead," it looked very long indeed.[26]

Even as the EDF worked through conventional channels of government, Congress was creating new ones. Public clamor for environmental action had reached the legislature, and bills that had been languishing in committee for several years suddenly came to the floor for consideration. Of these the most important was the National Environmental Policy Act (NEPA), which established citizens' rights to a clean environment and a process by which the government would assess the effects of its actions on the ecosystem (the controversial Section 102 of the bill).[27] There had been piecemeal commitments to environmental action during the previous decade, and Secretary of the Interior Udall had tried to make his agency into the government's conservation agency, but none of these actions had either the psychological or legal impact of NEPA. For the first time Congress committed itself to an ongoing study of human action in the environment and a policy of halting or reversing environmental degradation. Even without the guarantees of section 102, the environmentalists would have had an important weapon in NEPA; the addition of that section provided an unusual legal avenue for environmental concerns. It required that federal agencies assess the environmental impact of any proposed project and submit a statement of predicted changes for agency review. The adequacy of environmental statements, the nature of the proposed review, and the projects to which it applied have provided numerous chances for environmentalists to challenge projects and to gain access to the agencies' decision-making processes.[28]

Congress moved beyond the declaration of policy by providing the Council on Environmental Quality, and executive reorganization established the Environmental Protection Agency. The EPA, an amalgam of old offices from various departments, came into being with Reorganization Plan Number 3, effective 2 October 1970. More important for our purposes, responsibility for pesticide registration and regulation was transferred from the USDA's Pesticide Registration Division to the EPA in December, just as William D. Ruckelshaus took office as the first administrator. The change was a real shift in power, for whatever public constituency the EPA found, it would not be the current defenders of DDT. Both the NACA and its Congressional allies fought back, urging the administration to hold the line against immediate suspension of DDT (arguing that it would cause an increase in the use of more dangerous chemicals), and Congressman Jamie Whitten of Mississippi applied his own resources. Two months after the EPA received jurisdiction over pesticides, Whitten's subcommittee on agricultural appropriations received authority over environmental issues and consumer protection, giving the Mississippian power over EPA's budget.[29]

Neither the passage of NEPA, nor the formation of the Environmental Protection Agency, nor the reshuffling of responsibility meant that the battle against DDT was over, for EPA was no more willing than the USDA to take radical action. The EDF promptly went back to court in *EDF v. Ruckleshaus (I)* (there were two separate actions with the same principals), seeking review of Hardin's failure to cancel registration of DDT and to suspend use during cancellation proceedings. Early in January, Judge Bazelon and two colleagues ordered Ruckleshaus to cancel all uses of DDT immediately and remanded the case for decision on immediate suspension. Ruckleshaus complied, but, after a sixty-day review, refused to suspend registrations. DDT, he said, was not an imminent health hazard.[30]

By this time the environmentalists' patience was wearing thin. In July, the National Audubon Society listed, in a piece entitled "The Non-Ban on DDT—A Federal Phase-Out Fizzle," the actions that the environmentalists and the government had taken since the administration's promise to ban all but "essential uses" on DDT. The target date was only seven months off, it said, but "thus far industry appeals to USDA cancellation orders have substantiallly blocked the much-heralded DDT phase-out."[31] The operative word, though, is "blocked." The defenders of DDT were on the defensive, and although they could delay a ban on DDT, there seemed no way they could avoid it. Public opposition was stronger that ever before, and the courts had consistently ruled that the environmentalists had standing and had granted relief. Congress, by passing the NEPA and then taking pesticide regulation out of the Department of Agriculture, had further weakened the pesticide manufacturers hold on policy. The situation was almost the reverse of what it had been in 1967, when the ill-funded and very tiny band of pioneers had taken to the courts.

The court of appeals' order to cancel all uses of DDT set the stage for the final scientific confrontation over DDT. Under FIFRA, interested parties could appeal an order to cancel the registration (permanently ban use), and demand a hearing. Thirty-one formulators and manufacturers of DDT protested the order, and brought on the Consolidated DDT Hearing, which began in August 1971 and ran on into the following March. It was the most extended public discussion of the case for and against DDT ever held, as 125 expert witnesses and several attorneys produced over 9,000 pages of testimony in eighty days. Like the Madison hearing, it was conducted like a trial, with one side presenting its case, then the other, and each side having the chance to cross-examine its opponents' witnesses. The focus was the safety of DDT as determined by scientists. In other ways, though, the two hearings were much different; the issues, the legal position of the two sides, and even the composition of the coalitions had changed. In Wisconsin the environmentalists had been seeking a judgment against DDT and had brought the action; in Washington the defenders of DDT brought the action to protest a judgment already made. In Madison the statute under which the hearing took place had allowed full scope to both sides to develop their case. In Washington the legal issue was different; the question was the adequacy of labels for the uses of DDT under the cancellation order. Would the label directions, if followed, give adequate safeguards against harm to man and animal? Although the environmentalists fought in the Washington hearing on the broad grounds that any use of DDT would contaminate the ecosystem, they were in theory (and sometimes by the examiner) confined to a much narrower proposition, a circumstance that the Group Petitioners (the formulators and manufacturers) used to advantage. Government agencies also played a dominant role in Washington, which they had not done in Madison, where the industry contended with the private environmental group. Although the EDF worked closely with the Environmental Protection Agency, even suggesting witnesses and coordinating strategy, it did not direct the attack.

When Hearing Examiner Edmund Sweeney opened the hearing in August, in a hearing room in Alexandria, there was a large cast, even without the witnesses. Two Washington attorneys, Charles O'Connor and Robert Ackerly, represented twenty-seven companies, lumped together as Group Petitioners. Elliot C. Metcalfe and Raymond Fullerton represented the USDA, which joined the companies in protesting the cancellation. A firm that grew sweet peppers on the DelMarVa peninsula, H. P. Cannon, appeared to defend its use of DDT, as did other minor parties. On the other side was Blaine Fielding, a lawyer for the Environmental Protection Agency—the respondent-in-chief—and William A. Butler, the EDF's new attorney, represented a set of intervenors—the

EDF, the Sierra Club, the National Audubon Society, and the West Michigan Environmental Action Council.

The hearing began with Sweeney defining the issues—the adequacy of the label directions to prevent harm to the applicator, to animals, and to the ultimate consumer. The burden of proof, he said, rested on the petitioner registrants; they had to show that DDT was safe for the uses under question. (1–6)[32] The legal setting defined the strategy. The petitioners had to bring enough evidence to show that DDT was not harmful, a task they could accomplish only if they could undermine the basic case of the environmentalists, who would contend that the properties of DDT made discriminate use impossible. The hearing revolved, as it had in Madison, around environmental evidence but both sides also had to consider human health; the new cancer studies had made that a crucial point. The basic issues thus became the adequacy of the evidence of the environmental contamination (especially the analytical work), the significance of that evidence (thin eggshells in particular), the justification for extrapolating from the laboratory to the field, and the significance of animal tests for judgments about human safety.

Group Petitioners presented their case first, followed by the USDA, which supplemented and complemented their evidence. Robert Ackerly, representing the industry, stressed several themes in his opening statement. There was, he said, a long history of DDT use, free from illness or death; DDT had made many contributions to human health and to agriculture and it was still needed. The case against the chemical was, he went on, weak. There was analytical confusion between DDT and PCBs, which cast considerable doubt on the claims of worldwide contamination. Even if DDT was spread throughout the world, the level was low and even now declining, and the significance of the residues and their effects on living things was unclear. Given this sketchy case, DDT's outstanding safety record, and the demonstrated need for the chemical, the balance of costs and benefits clearly dictated continued registration of the uses at issue.(14-29)

An element in the defense of DDT was its relation to

human health—its use in controlling the insect vectors of disease and the evidence linking it to human pathology—but it was hard to make a case that DDT was needed in the U. S. to control disease. Although the witnesses could cite uses of DDT to halt malaria in many countries, they could not show that it was essential in the United States. Indeed, Jesse L. Steinfeld, surgeon general of the Public Health Service, admitted that he did not know "of a specific instance where it might have been used in the last 20 years" in the United States.[33] Still, a ban on DDT in America would, these witnesses feared, lead to bans in countries that desperately needed DDT.[34]

Group Petitioners and the USDA also sought to show that there was no real danger to man from exposure to DDT. The testimony stressed three points: first, there were studies that indicated that DDT was not harmful; second, extrapolation from animal doses at very high levels to humans exposed to very low ones was difficult and unrealiable; third, the demographic evidence did not show a link between DDT and human cancer. To show that DDT was safe, the petitioners and the USDA called both Wayland Hayes and his assistant, Dr. Edward Laws, now chief resident in neurosurgery at Johns Hopkins, to go over the studies they had done on workers and convicts. In cross-examination Fielding and Butler went over the same ground Yannacone had covered in Madison.[35] The USDA also offered testimony on the lack of connection between DDT and cancer, bringing several officials of community health programs to testify that they had found no pathological conditions among persons using spray equipment. The agency, clearly, wanted to calm the public; one witness stated frankly that his work should debunk "the carefully cultivated myth that DDT is highly toxic to man . . . and the view now widely held by the general public that contact with DDT results inevitably and promptly in death from cancer."(2075-2076) In attacking the use of animal tests, the witnesses did not completely disregard them. Dr. Ted Loomis, a pharmacologist from the state of Washington, admitted that there was a possibility that in very high doses

DDT might be tumorgenic. Since the dose-response relationship for potential carcinogenicity was not well worked out,
though, he preferred to place more weight on human data, if
that was sufficiently extensive, than on evidence from animals.(664) The weight of evidence, he said under cross-
examination, indicated that DDT at current levels in man was
not carcinogenic.(701)

The medical evidence was not as definitive as the defenders
of DDT would have liked. The USDA and the EPA offered
as neutral witnesses two officials of the International Agency
for Research on Cancer, Drs. Lorenzo Tomatis and Boyd
Higginson. Neither said that DDT was clearly harmless to
man or that the evidence was sufficient to clear it. On the
contrary, they were concerned about the implications of the
animal tests. Tomatis, a physician and oncologist, thought
that there was no evidence to show that DDT was carcinogenic in man, but it was definitely carcinogenic in mice.
Although not definitive, this evidence did indicate possible
danger. Higginson, director of the agency, said that the
tumors in mice should cause grave concern, though there was
no complete correlation between man and animal. "[A]nimal
experimentation in relation to compounds to be utilized in the
future is one of the few protective lines we have."(1458) Still,
in view of DDT's record, he thought there was no reason not
to continue prudent, limited use. DDT was certainly not a
potent carcinogen; although the epidemiological data on man
was not sufficient to settle whether the chemical was or was
not a weak carcinogen, it certained ruled out its being a strong
carcinogen.

The medical situation became even more complicated with
the appearance of Dr. Umberto Saffiotti, associate science
director for carcinogenesis of the National Cancer Center,
part of the National Institutes of Health, who discussed for
Respondent EPA both the general characteristics of carcinogenesis and the results and significance of the animal tests.
In the case of DDT, he said, the evidence was clear. In two
strains of inbred mice, the chemical had produced liver
tumors at rates significantly higher than those in controls, and

tests on two other strains, one inbred the other not, had confirmed these results. Other studies—on rats, mice, and trout—had flawed research designs, but added support to the evidence. Saffiotti also offered a critique of studies that purported to show that there was no danger from DDT. Negative studies, he said, had, without exception, inadequate experimental designs—lack of controls, insufficient numbers, or other factors that invalidated the results. Work on hamsters was suspect for another reason; well-known carcinogens had failed to produce positive results in this species. What, he was asked, about Laws's paper, which indicated that DDT might have antitumorogenic properties? It was, he said, so bad that it "discredits not only the author, but the journal that accepted it for publication."(4075) The experimental design was bad, it did not even deal with the induction of tumors, but with transplanted tumors, and the effects it outlined were so small they were insignificant.[36]

The USDA countered by bringing to the stand Alice Ottobomi, a biochemist who had studied the effects of DDT on rats and beagles for the National Institute of Environmental Health Sciences. She had not, she said, found adverse effects from exposure to DDT. Three generations of rats and two of beagles had shown no decrease in fertility or viability of offspring and autopsies had shown no liver tumors. During cross-examination, EPA tried to show that her work had not been conducted under the same conditions as the studies Saffiotti had cited, and was not comparable. Rats in the reproduction studies had been killed after the weaning of the second litter and those in the study on age and fertility at various points during the normal lifetime of two years.(4124–4126) Nor had all the animals been carried through the experiment; rats that did not produce young were sacrificed, and dogs carried through the lifetime had been selected by a nonrandom process. Her work was a toxicity study, not a test of carcinogenesis or mutagenicity.

The medical evidence was the clearest instance in which the scientists were unable to answer a question the public and the agencies wanted answered: was DDT carcinogenic to man?

The answer depended, not on the experimental results, but on their interpretation, a matter that could not be resolved by any simple process. There were two opposed schools of thought. One believed that tests of animals exposed to high levels of a chemical were a good, if imperfect, indication of carcinogenicity in man. The other relied more heavily on the evidence of human cancers in exposed human populations, a test that the first group believed was not adequate. Neither school of thought, to complicate matters, condemned the other; each saw some good in the work that its opponents did, neither could be convinced, though, that as much reliance should be placed on the opponents' position as its supporters thought should be.

The USDA bore the main burden of showing that the environmentalists' contentions about DDT contamination were false. Its case, though, was inconclusive. It did not seriously challenge the base of scientific studies on which the EDF and the EPA case would clearly rely, but managed instead only a variety of specific criticism of particular studies. On the thin-eggshell controversy, for example, Thomas Maren, a pharmacologist from the University of Florida Medical College, said that he had been unable to reproduce the work of Peakall and Bitman on the inhibition of the carbonic anhydrase reaction by DDT. He admitted, under questioning by Sweeney, though, that there were "at least ten other reactions" that would produce thin eggshells, and under cross-examination he said that he did not rule out the reaction of DDE on the formation of eggshells, only the reaction through that particular chemical route.(2568-2564) Kenneth L. Davison, a research physiologist with the Animal Science Research Division of the USDA, testified that he had found no effect on white leghorns when DDT was added to their diet, but stressed that there were wide differences between various species.

At other points USDA witnesses admitted parts of the environmentalists' charges. Gary Booth, project director of pesticide pollution in the Illinois Natural History Survey, discussed the concentration of DDT by organisms in con-

taminated water and said that if there was an organism that
was a good "sponge" for the chemical, the accumulation in
the upper trophic levels "will become catastrophic."(3053)
George Harvey, a scientist from Woods Hole Marine Biological Laboratory, admitted on cross-examination that although
the signals of PCBs and DDT overlapped, the distinctive
peaks of the Arachlors (PCBs) made it easy to separate the
compounds.(3034-3101) Ned Bayley, director of science and
education for the USDA said that the department recognized
the harmful effects of pesticides and was committed to
reducing their use as much as possible.(3004-3125)[37] The
USDA was clearly shifting ground. In 1967 it had defended
both DDT and the agency's autonomy from public scrutiny.
Now, with authority transferred to another agency, the
scientific case becoming more and more definite, and public
opinion shifting against DDT, it conducted a minimal defense. DDT, it argued, was still needed for some tasks, but it
would soon be replaced.

   The USDA ended with Bayley's plea for a few more years
to phase out the chemical, and the EPA began its presentation. Blaine Fielding began by outlining the agency's position:
DDT was not needed and it was dangerous. EPA would
show that it was DDT, not PCBs, that were being found in
such abundance in the environment, and that the chemical
was persistent, mobile, and concentrated by biological organisms. It would bring witnesses to show that DDT caused
damage at all levels in aquatic ecosystems and that it affected
commercial fishing. A variety of studies from field and
laboratory would demonstrate DDT's harmful effects on
birds. With regard to human health, EPA would show that
the chemical had tumorogenic and and mutagenic properties
that were relevant to an assessment of its danger to man.
Finally, he said, other methods of insect control could replace
DDT for even the so-called essential uses. The argument was
the one EDF had made in Madison, and EPA would even call
on some of the same witnesses. It omitted only the discussion
of the ecosystem, stressed more strongly the effect of DDT on
aquatic life (particularly the commercial dimensions of the
problem), and added the issue of human health.

EPA began with three witnesses—two from the Department of the Interior's research laboratories, one from the FDA—who testified that DDT could analytically be distinguished from the PCBs. Then it introduced scientists—from university laboratories, environmental agencies, and even the USDA's Agricultural Research Service—to discuss DDT's properties in the environment. DDT, they said, constituted a burden on ecosystems. Its persistence, its movement through the air and water, and its bioconcentration made it dangerous far from the point of application and for many years after it was applied. The petitioners cross-examined several of these witnesses at length, raising various alternatives to their interpretation of the data, attempting to clear DDT by implicating other environmental contaminants. For example, biologist Tony Peterle, who testified about DDT contamination of Antarctic fauna, faced a barrage of questions about the conditions under which he collected samples, various possible sources of contamination, and possible confusion in the analysis.[38]

The EPA moved on to aquatic ecosystems, bringing several scientists to the stand, including Robert Reinert, a witness the EDF had planned to present in Madison, and Kenneth Macek, who had appeared at that hearing. Several witnesses appeared to discuss the effects of DDT on various species. Two, from the Patuxent Wildlife Research Center, were particularly important. William H. Stickel's testimony was similar to that given by Hickey and Wallace in Madison about the effects of sprays for Dutch elm disease, and Robert G. Heath's work was an expansion of the studies reported in Madison by Lucille F. Stickel. Both, though, had more evidence than had been available in 1968. Cross-examination was vigorous, for the combination of studies was important for the environmentalists' case. Lawyers for the petitioners asked Stickel about the studies of migratory hawks done at Hawk Mountain, Pennsylvania, and about the Audubon's Christmas bird counts, although Stickel repeatedly denied the usefulness of these studies in assessing continental bird populations and refused to draw any conclusions about these populations as a whole from any evidence. No one, he thought, knew enough to say whether those particular species were increasing or decreasing. In cross-examining Heath, attorneys raised the possibility that other factors, including stress from caging or mercury compounds, might be causing thin eggshells, asked him about studies on other species that did not show the phenomenon, and objected to his experimental designs.

The lawyers were not alone in questioning these witnesses sharply; Sweeney often pressed them. One example from Stickel's cross-examination will suffice.

> *Sweeney:* Can you give an answer, yes or no? He is asking you a question. You say yes or no. If he wants further, he will ask you further.
>
> *The Witness:* Sometimes the answer yes or no gives a misleading answer.
>
> *Sweeney:* We'll strike that from the record. I'll instruct this witness to answer the question yes or no when he can do it. I instruct him to do it now.(4419)

His rough handling of the two scientists caused a minor rebellion. The day after Heath's cross-examination, Alan Kirk, deputy general counsel for the EPA, appeared to say that the scientists from Patuxent, under orders from their supervisor, Eugene Dustman, would not come to testify. Dustman, he said, was also petitioning the Department of the Interior to keep all witnesses from the agency out and was asking that the Fish and Wildlife Service not be asked to appear at any hearings held under FIFRA. Sweeney, he alleged, was making off-the-record remarks impugning the scientists, holding them up to ridicule, and was exhibiting a strong bias in the case. Fielding, who also represented EPA, though as counsel for the respondent, came to Sweeney's defense. He also launched an attack on Yannacone's handling of the Wisconsin case, which he characterized as a "circus." Carefully disassociating Butler, who was not present, from the "then counsel for the Environmental Defense Fund," Fielding pointed to the awful consequences of Yannacone's cross-examination, which had so intimidated many scientists that EPA now found it difficult to get witnesses to appear.(4588) Group Petitioners joined in. O'Conner said that Sweeney had been conducting the hearings in accord with the rules of procedure. Heath, he claimed, had been evasive and uncooperative and had appeared without the data to support his work. The next day, as the dispute dragged on, he suggested that the EPA was simply looking for an excuse not to bring its witnesses.(4606-4608)

The blowup over Sweeney's conduct of the hearing was not a complete surprise. There had been rumors even before the hearing had begun that the EPA had wanted another examiner for the case, and quite early EDF people had begun to grumble in private that Sweeney was treating Butler like a second-class citizen, allowing him to expand on the EPA's cross-examination only when it suited him, not enforcing the rules of evidence with regard to competency, and loading the record with "junk." Sweeney's short way with dilatory witnesses had also drawn comment, but Dustman's complaints went well beyond the earlier charges, for they raised questions about Sweeney's competence and fairness. The examiner certainly was unsympathetic with the attempts by Stickel and Heath to qualify their answers, but his impatience also extended to USDA and Group Petitioners witnesses who sought to avoid yes-or-no answers. The incident was smoothed over; with assurances from Sweeney and the EPA that the scientists would be treated fairly, Dustman allowed his subordinates to return to the hearing.

After two more biologists from Patuxent testified on feeding studies done with DDT, the EPA turned to field observation on brown pelicans, which turned out to be a controversial subject. James O. Keith, chief of the section of pesticides and wildlife ecology at the Department of the Interior's Denver Wildlife Research Center testified about conditions on Anacapa Island, just off the coast of California. In April 1969, he said, biologists visiting the pelican rookeries had found only nineteen normal-looking eggs (all of which proved to be thin-shelled). The others were abnormal "col-

lapsed, dehydrated eggs, [which] looked like roasted marsh-
mallows. . . ."(4823) Tests had eliminated both mercury and
lead, two obvious possibilities, and Keith said that DDT was
the culprit. No other compounds "even approached this kind
of effectiveness in thinning eggshells. . . ."(4835) The petitioners vigorously disputed these conclusions, and cross-
examination raised a number of other things that, lawyers
felt, might be responsible for the problem. Outside the
hearing room J. Gordon Edwards, an entomologist who was
one of the most active defenders of DDT, pointed to the Santa
Barbara oil spill and the subsequent disturbances in the area,
including helicopter landings on the island, as more likely
causes. In other areas, he said, pelicans were reproducing, so
DDT clearly was not responsible.

The peregrine falcon followed the pelicans as the EPA
brought Joseph Hickey, Tom Cade, director of the Cornell
Laboratory of Ornithology, and David Peakall, who had been
slated to appear in Madison, to testify on field and laboratory
studies on the hawk. Cade, who had become convinced of
DDT's role in the thinning eggshells at Hickey's 1965 conference, testified about the reproductive failure among peregrines nesting on the Arctic slope of the Brooks Range in
Alaska. Peakall testified on his work with enzyme induction
and was, oddly, excused without cross-examination. Hickey
repeated the testimony he had given in Madison on eggshells
and added other work on more species. The data base was
much broader—20,654 eggs laid before 1947 and 3,004
after—and it showed that there was a decrease in average shell
thickness in the eggs of virtually all fish-eating birds in North
America. Thirty-nine species were affected and ones common
to Europe and America showed the same problem on both
continents. "DDT is an extremely important cause of this
wide-spread disease, that evidence is indeed overwhelming,"
he said.(5021)

Cross-examination followed by-now-familiar lines. Attorneys challenged the research methods and design, probed for
more data, and suggested alternative explanations for the
environmental scientists' conclusions. They were more suc-

cessful than McLean had been—a testimony to the preparation the industry had given to the Washington hearing—but
the method of attack was weak. Industry lawyers could not
appeal to an established body of theory or to scientists who
had conducted research in the area. They could only raise
hypothetical objections, ask for more data and introduce
studies that seemed to contradict some aspect of the environmental scientists' work. They could not examine the material
in depth, show that apparent contradictions were real, or
mount a serious counterattack.

Finally, the EPA introduced testimony that DDT could be
replaced without ruining the cotton farmers, the principal
users. Joseph C. Headley, professor of agricultural economics
at the University of Missouri, discussed the expense of
substitute insecticides and USDA studies of their use. He
concluded that there would be no appreciable economic
difference if DDT were banned from the cotton fields. The
pesticide manufacturers raised a familiar objection—that substitutes were more dangerous to the applicator. The or-
ganophosphates, in particular, could cause serious illness or
death if mishandled. Headley defended the use of substitutes,
pointing out that the compounds could be used safely if
farmers and applicators followed proper safety precautions.[39]

Another controversy over Sweeney's competence interrupted this line of testimony. Immediately after Headley's
cross-examination, O'Connor brought in an article that had
just appeared in *Science:* "DDT: In Field and Courtroom a
Persistent Pesticide Lives On."[40] It recounted the progress of
the DDT battle from Hardin's cancellation of registrations in
November, 1969 and described the hearings. "Attorneys for
the EPA, as well as others who have taken part in these
marathon proceedings," the article said, "say that Sweeney
has often failed to act impartially, that he seems to lean
toward the industry's view-point, and that he has 'insulted'
some witnesses." It quoted one unnamed EPA attorney as
saying that there had been friction between the agency and the
examiner "from the start." Another (also anonymous) said
that Sweeney treated witnesses inequitably, and a third that,

though Sweeney was "doing the best job he can," he "doesn't understand scientists or scientific methodology." "EPA officials," it concluded, "seem more disposed toward heeding the advice of a special panel of scientists who prepared a report on DDT for the agency at the industry's request," than did Sweeney, who seemed to have "made up his mind a long time ago." O'Connor indignantly charged that the EPA was using the press to question the "integrity of this administrative forum and the objectivity and capacity of this tribunal."(6233)

Sweeney adjourned the hearing for the day, until Fielding could answer questions about the article. The following morning the attorney said that EPA had "no question as to your neutrality in these proceedings,"(6238) and asked why O'Connor had found it necessary to bring in the article in the first place. The EPA attorneys whose comments appeared here were, he said, nonpolicy staff, and, "under direct orders from [his] superior," he refused Sweeney's demands that he name them. At this Sweeney adjourned the hearing for another day to consider his response, and the proceedings resumed only after the offending article had been read into the record.

The EPA then continued its case against DDT for cotton insect control. The witnesses admitted that DDT had been useful on cotton in earlier years, but pointed to the development of resistance among major pests, the sudden (and in some cases disastrous) increase in insects that had been only minor pests before, the environmental effects of continued DDT use, and the superiority of integrated controls. Cross-examination stressed the different conditions in California and the Deep South and the witnesses' lack of experience in the latter area.

Before the hearing started, Montrose Chemical Company had requested an independent review of the data. The report was ready now, and Fielding attempted to introduce it as evidence. Sweeney, after hearing objections from both Group Petitioners and the USDA, rejected the proffer, saying that the scientific advisory report and the hearing should be separate. The EPA could, he said, offer witnesses, but not the document. At this point the EDF offered a motion for a directed verdict. Sweeney rejected it without even reading it. He knew, he told Butler, that it was offered solely for the record, and he did not feel like wasting his time. On this note the hearings adjourned for Christmas.[41]

On 4 January 1972, the fifty-eighth day of the consolidated DDT hearing, William Butler, representing the Environmental Defense Fund, the National Audubon Society, the Sierra Club, and the West Michigan Environmental Action Council, began the presentation of the environmentalists' evidence against the continued registration of DDT. The case, he said, would complement, not duplicate, that presented by the Environmental Protection Agency. "[C]onclusive evidence" exists "that there is substantial environmental harm caused by DDT to nontarget species of organisms" and "presumptive, if not conclusive, evidence that DDT is a carcinogen and a mutagen. We feel that it's clear that DDT is a nerve toxin, and an enzyme inducer and we would feel that burden of proof is upon those who would use DDT in light of these findings, and we would submit that this burden of proof has not been met."(6521) The testimony that followed was, however, familiar, for the EDF was simply "mopping up" behind the EPA's array of witnesses.

After predictable testimony on carcinogenesis and the criteria for judging the dangers of a chemical to man (and equally predictable cross-examination), the EDF brought Robert Risebrough to the stand. Some of the testimony—on the mobility and bioconcentration of DDT—was quite familiar, but Risebrough also attempted to answer some of the charges being made outside the courtroom. To head off criticism of the pelican studies on Anacapa Island, he discussed the precautions scientists had taken to avoid disturbing the nesting sites, and Butler attempted to introduce J. Gordon Edwards's press release so Risebrough could rebut it.(6848) Risebrough also attempted to defuse some of the testimony given by USDA witnesses by admitting that the exact mechanism of eggshell thinning was still unclear. Work done recently, he said, had ruled out the inhibition of the calcium anhydrase enzyme as the sole cause, and now the inhibition of other substances involved in membrane transport seemed a more likely explanation. It was certain, though, that thinning was occurring in the wild and that DDT was responsible. (6851–6873)

The EDF brought a series of witnesses to discuss various aspects of DDT's effects on man and wildlife. One discussed its possible role in genetic changes in *Drosophila* (fruit flies) in the western United States. Another, George Woodwell, reported on the composition and abundance of species in the various trophic layers in a marsh, and an environmental scientist from Virginia challenged the effectiveness of the quarantine against the gypsy moth and the use of DDT on trailers. Samuel Epstein, a professor of environmental health and human ecology, testified about the studies on the carcinogenicity of DDT (largely a summary of earlier testimony on the subject). A scientist discussed DDT's and PCBs' effects on phytoplankton, and an insect toxicologist from Texas, Frederick Plapp, discussed resistance to pesticides and the use of integrated controls.[42]

With the EDF's testimony in the record, the petitioners offered their rebuttal case, contending that the chemical did not spread through the environment and that it was essential for cotton production in the South. They mounted a new assault on the evidence that DDT or its metabolites caused thin eggshells, bringing in new feeding studies, a new critique of the methodology of the environmentalists' work, and evidence that wildlife populations were flourishing. Finally, they brought more witnesses to testify that DDT was not a carcinogen. Like the EDF's case, much of this was plowing already well-worked ground. The only really new development, testimony on the statistical evidence linking DDT and eggshell thinning, was given by Bruce C. Switzer, a graduate research associate at the University of Alberta and a doctoral candidate in zoology and food science, who had been working since 1969 on the effect of insecticides on the reproduction physiology of common terns and had found no evidence of shell thinning. Hickey and Risebrough, he said, used "inap-

propriate statistical analytical techniques," that, by distorting
the resulting correlations, had led them to conclusions that
could not correctly be drawn from the data. They had
assumed a normal distribution of the variables—that is, a
random distribution of the data about the mean, the distribution that would produce the familiar bell-shaped (Gaussian
distribution) curve. Shell thickness, though, did not vary in
this fashion, and it called for a different statistical technique.
The petitioners also recalled Risebrough for more cross-
examination on this point, seeking to show that his work was
invalid because of the choice of statistical treatments for the
data. He vigorously denied this, though he did admit that, in
some cases, a different statistical treatment would have been
"more appropriate."

Both direct examination and cross-examination were nota-
ble more for their attempts to bring out points favorable to
one side or the other than for their exposition of the problem,
and the exchanges suggest serious limitations on the use of
scientific testimony. The significance of the statistical tests
was lost and the basic thread of argument discarded as each
side tried to discredit the other. Although the medical witnesses had clear commitments to different methods of testing as
the best means of assessing unknown, and unknowable by
direct measurement, hazards to humans, and the criteria had
been clearly developed by the testimony of various experts, in
the argument over the statistical evidence the problem was
more technical and was not clearly developed. The environmentalists claimed that the problem was a minor one,
whereas the industry and USDA tried to show that the entire
edifice of the environmentalists' case rested on it. Neither
Sweeney nor the administrator of the EPA, who made the
final decision, met the issue squarely. Sweeney generally
rejected the environmental argument about eggshells on the
grounds that there was no direct connection between the uses
at issue and thin shells; the administrator weighed Switzer's
credentials against those of the environmental scientists and
found them wanting—a procedure that may have made sense
for a public official but did no credit to his grasp of science.

Sweeney, even in the hearing room, was quite impatient with the testimony on statistical variations, and his actions led the EDF to condemn these pages of the transcript(8506–8552) as "perhaps the most scandalous in the hearing for what they reveal as the hearing examiner's hatred and contempt for the EDF."[43] At one point Sweeney rejected a list of data offered by Risebrough on the grounds that it was a computer printout, and threatened to strike all Risebrough's testimony because; when asked for data, he had brought "a bunch of computer printouts," which only a computer could understand.(8523) He rejected Risebrough's explanation that these were simply tabulations of data, the kind he normally did with a typewriter. After one particularly vigorous passage with Butler, Sweeney remarked: "You know, Mr. Butler, if you were six years old, I'd know how to handle you."(8542–8543)

Oral arguments followed the rebuttal case, with the Group Petitioners and USDA presenting a solid front. As far as human health went, they said, DDT was certainly safe. The testimony of "some of the most distinguished world-renowned professionals who have ever lived,"(9094) "medical authorities whose responsibility it is to make determinations of human health hazards . . ."(8991) agreed that the chemical did not cause cancer. There was, in addition, the experience of DDT use. "We find it inconceivable that Respondent and Intervenor EDF have, in effect, placed 72 mice ahead of a quarter of a century of DDT's usage unsurpassed in the history of man in terms of safety as a chemical pesticide. . . ."(9051) The environmentalists argument was, they said, irrelevant and inaccurate. Since the environmentalists could find no harm to wildlife in the South, they had resorted to "this shibboleth of mobility. . . ."(8999) Aquatic injury, which the environmentalists had stressed, was "virtually without exception the result of direct local use in and around aquatic areas."(9004)

The real issue, O'Connor said, was the objectivity of the scientists. He listed alleged gross breaches of that objectivity, including the testimony of chromosomal change in fruit flies and Woodwell's figures for DDT contamination of the Long Island marsh (the result, it was claimed, of the dumping of DDT). Hard judgments, he said, must be made, "if scientific truth is going to be protected." Hickey had improperly used statistical techniques in his study of eggshells and discarded evidence, such as the Hawk Mountain Sanctuary counts ("the best evidence"); Audubon had also used these figures, he said, until they ceased to serve its purposes. The pesticide industry, though, had not been sensitive to the question and had not sponsored studies to come up with contrary data, so they had been forced to rely on cross-examination,(8977) what Yannacone would have called the "crucible of cross-examination."

DDT's opponents made a more formal division of labor. The EPA dealt with the legal issues and human health, whereas Butler discussed the environment and the necessity of continued DDT use. The burden of proof was on the Group Petitioners and the USDA, EPA's lawyer said, citing *EDF v. Ruckelshaus* on the interpretation of FIFRA.(9079) The petitioners had not, he went on, shown proof and had not controverted the evidence offered. With regard to human health, though, he did not take a hard line. He admitted the difficulties of extrapolating from animal studies to humans, but said that it was better to be cautious than run grave risks to all. As for the demographic data, even the best authorities, including Jesse Steinfeld, had admitted that it was weak. The lack of evidence, he pointed out, did not mean that DDT was safe, but as Steinfeld had admitted, it only showed that the chemical was not an imminent hazard. Butler went over equally familiar ground. DDT was persistent and mobile, he said, citing the various witnesses who had appeared to testify on these properties. Midway through this presentation Sweeney interrupted. "Are you—is it your position that no label can be written for any use here that will satisfy FIFRA?" Butler: "Yes."(9118) Butler then rebutted the rebuttal witnesses, showing that Switzer's own treatment of the eggshell data gave only slightly different results, pointing to places where the petitioners' witnesses had corroborated the environmentalists' testimony, and denying many of their other points.

Ackerly for the Group Petitioners and finally Fullerton for the USDA replied to that, and then the hearing ended in a flurry of technical motions, admission of evidence, and other material. After eighty days, over 100 witnesses, and a pile of documents, the hearing was over. The various parties submitted briefs analyzing the evidence, showing how the legal issues should be interpreted, and making a final plea for their cases.

Qualified Victory

THE EPA HEARING was the last major confrontation over the scientific evidence of DDT's effects on man and in the environment, and even as it went on the battle over persistent pesticides and the fight for a cleaner environment swept past it. Scientists who came to testify in Sweeney's hearing room had, in some cases, already appeared before Congressional committees considering sweeping changes in regulation, or they had helped put together cases against other persistent pesticides. The environmentalists had already moved against dieldrin, aldrin, and Mirex, were preparing cases against other chemicals, and were involved in a score of projects, from the Alaska pipeline to international whaling. An official end to DDT, though, was still important to the EDF and to other environmentalists, and not just for its own sake. The issue was a test of the EPA's willingness to take strong action; a legal victory would be an important legal and psychological boost for the cause. In the spring of 1972, then, the storm center was the office of William D. Ruckelshaus, administrator of the EPA. What would he make of the evidence? He had three documents to consider. There was the record, which was an enormous body of evidence, and two opinions on the evidence. One was Sweeney's, the other was the report of the advisory commission appointed at the request of Montrose Chemical Company when the cancellation came down (the Hilton Commission). This group had finished its work before the hearing ended and, although its opinions had no legal force, they could be an important factor, for the members were respected scientists. What would Ruckelshaus decide and how would he justify his decision?

He had first to deal with the examiner's decision, delivered a month after the hearing ended. Sweeney had, not surprisingly, affirmed the industry's arguments. The basic problems, he believed, were to weigh the benefits and risks of DDT use and to determine if the labels in use gave adequate directions to "prevent all unreasonable injury." A review of DDT's use left "little question of the far-ranging public health and welfare benefits from DDT" and the important, if less dramatic, benefits to food production. There was no evidence that DDT caused problems in man and "[t]hose that would ban all use of DDT because of the possibility of some danger to man, the evidence of which is said to consist of the results of a few experiments with animals, would do well to compare such skimpy evidence of risk with the well-documented proof of the benefits which DDT has bestowed on mankind."[1] The evidence on DDT's carcinogenicity and mutagenicity in other forms of life he rejected on the grounds that it could not validly be used to predict the effects in humans; the best evidence, he thought, had been offered by such authorities as the surgeon general, who had found no evidence of harm to man. The environmentalists' testimony on DDT's effects in the ecosystem he found suspect on several grounds: some parts were conflicting; others, including evidence of biomagnification, were weak; there were difficulties in extrapolating from the laboratory to the field; and much of the testimony was irrelevant to the issue at hand. There was no evidence on birds, for example, that "focused its direct thrust on damage . . . by the uses of DDT. that are permitted under the registrations in question." DDT, he concluded, was not a hazard to man and the uses involved would not have a deleterious effect on wildlife.

Sweeney's opinion was in sharp contrast to Van Susteren's belated decision, filed a year after the Wisconsin hearing ended (though it was dated May 1969), and it is instructive to compare the two.[2] The Wisconsin hearing examiner, working from similar evidence, found that DDT was mobile, stored in fat, and present in all levels of the food chain. The bioconcentration of DDT, he said, made it impossible to establish safe levels 'of exposure or safe methods of use if the chemical was toxic, and the evidence indicated that it was.

From the combination of work on DDT's effects in the nervous system and its use as a rodenticide "the only valid permissible inference is that DDT in small dosages has a harmful residual effect on the mammalian nervous system." It was also clear, from laboratory and field studies, that DDT was having an effect in the environment—it was thinning eggshells in certain susceptible species. Unlike Sweeney, Van Susteren did not consider the evidence of DDT's past benefits. "Without doubt," he said, "DDT has provided enormous economic benefits, but economic benefits are not an issue or part of any issue in this case.[3] DDT, he ruled, was a pollutant within the meaning of the Wisconsin statute. (The pesticide's defenders objected to the opinion on the ground that the question was moot—legislative action had already ended DDT use in Wisconsin—but they were unsuccessful in getting it set aside.[4])

The administrator did not consider Van Susteren's report, but he did have the reports of his own select commission (the Hilton Commission) and that of the Secretary of Health, Education and Welfare (the Mrak Commission). Both were much more critical of DDT and more concerned about the possibility of unknown dangers than was Sweeney. The Hilton Commission had found that, despite the decrease in the use of DDT over the last decade, quantities in the soil and water had not diminished. DDT and its metabolites spread through and persisted in the global ecosystem and, because organisms concentrated the residues, the chemical threatened at least some species. Although the commission did not believe that DDT had been proven to be a threat to man—it rated the dangers of tumorogenesis and carcinogenesis as "low"—it recommended an accelerated reduction of DDT use and the development of alternative methods of control.[5] The Mrak Commission had come to similar conclusions in 1969. Although it did not regard the evidence on DDT's effects in the environment as decisive—it said that there was far too little information available on natural populations and too little research underway—it did believe that persistent pesticides caused at least some harm and recommended their elimination from general use in the next two years. On carcinogenesis it said that suspicion had been aroused by the National Cancer Institute tests; it should be dispelled by further research.[6]

Sweeney's decision stood alone, and Ruckelshaus followed the more beaten track. In June he overruled Sweeney and banned the remaining uses of DDT on crops, allowing only quarantine and public health uses and manufacture for export. The examiner's decision could not stand, he said, it was not based on a compelling choice of relevant testimony or a good assessment of the credibility of the witnesses. DDT was a persistent toxicant that was uncontrollable when released into the environment. Evidence "compellingly demonstrates the adverse impact . . . on fish and wildlife." With regard to human health there were no adequate epidemiological studies on man, and animal tests showed that the chemical "should be considered a potential carcinogen." Since there were effective and environmentally safer substitutes, the administrator ruled, all DDT use would end in six months, which would allow an orderly transition to other pesticides and adequate education in their safe handling.[7]

Although the decision in June marked the end of general use of DDT in this country, it did not mark the end of the fight over the chemical; industry and certain users continued a rear-guard action. The companies sought a judicial over-throw of the order, and the administrator did not get final vindication until December 1973 when the court of appeals for the District of Columbia ruled against the industry (The EDF also appealed, seeking to end even the manufacture and export of the banned chemical). There were also petitions for the emergency use of DDT. In 1973 and again in 1974 the EPA allowed its use against the pea leaf weevil in Washington and Idaho, and in 1974 it let the forest service spray thousands of acres in the Northwest to stop an infestation of the Douglas fir tussock moth.[8] Congressmen have also used their power to press this case on the EPA. Congress ordered the agency to review its decision against DDT and has threatened at various times to pass a legislative veto over the agency's decisions.

It is unlikely that the arguments, lawsuits, and lobbying will vanish. The relative peace of the pre-*Silent Spring* era was grounded in a widespread, if unspoken, consensus about values and dangers. Americans assumed that science was good, that chemicals were necessary, that their use would be governed by the experts, that these experts could be trusted, and that the side-effects of chemical use would be negligible. All of these beliefs have lost their hold and, in addition, many Americans have come to place new and higher values on wilderness, wildlife, and nature. At the same time, though, they have not abandoned the ideal of the good life based on high and rising consumption of energy and the use of new technology. The result has been laws, regulations, and agencies to implement the new ideals, which were in intense conflict with the older, still potent ones. Worse, the effects of environmental action bear unequally on different parts and classes of society. What appears a minor environmental improvement or a necessary step in a long-range plan may be a major threat to the short-term livelihood of a small group or a particular industry. Farmers, ranchers, and foresters have spent the last thirty years adapting their production and output to conditions set by the availability and wide use of inexpensive chemicals, and they cannot adjust overnight, even if they wish to. Some environmental projects threaten, or seem to threaten, jobs or the entire economic development of a region. The environmental issue has assumed in some places the aspect of a class war—rich against poor, locals against outside do-gooders. It is not clear what the outcome of this struggle will be; it is only possible to indicate the major effects of the DDT case on scientists, citizens, and public policy.

The first casualty of the battle over DDT was the old system of federal regulation of pesticides and the interest-group control over that policy that had been established in the post-World War II period. *Silent Spring*, and then the environmental movement, brought public and Congressional scrutiny of the process by which the USDA registered pesticides, and groups whose interests were affected by agricultural chemicals began to agitate for a voice in the formation of policy. Congress's initial response, after *Silent Spring*, had been the inadequate interdepartmental agreement of 1964.[9] Then in 1969–1970, there came the National Environmental Policy Act, the Council on Environmental Quality, the Environmental Protection Agency, and the transfer of much of the USDA's authority over pesticides to the new agency. The next year Congress began consideration of two bills, S. 745 and S. 660, which would revise the Federal Insecticide, Fungicide, and Rodenticide Act of 1947, the first major changes in that legislation since its passage. Both bills significantly added to federal power over the use of pesticides. They included control over intrastate acts, provisions for the recall of pesticides, stop–sale orders, and civil penalties, and increased criminal penalties and requirements for manufacturing and applying the chemicals.

The fight over the FIFRA revision was largely on agricultural matters. Some witnesses mentioned the risk of cancer, but most of the testimony, aside from that by environmental scientists, revolved around the cost of regulation to the farmer, the safety and availability of substitute compounds, and the effects of the legislation on southern cotton farmers. The Congressmen fully shared this concern; several appeared convinced that they faced a group of wild-eyed fanatics bent on doing away with people for the benefit of birds. The same concern was evident in the hearings the following year, as Congress continued the debate. By 1972, though, the focus had narrowed to specific provisions of the act and their effect on manufacturers and users.[10] The industry and cotton farmers seem to have accepted a new FIFRA as inevitable and to have spent their time making it as palatable as possible. Late in October, Congress passed a bill, but each succeeding year has seen a continuing struggle over the extension of the act and its provisions. The lack of consensus and strength of the opposing interests has left FIFRA, and pesticide policy, in flux.

The same ambiguous results are apparent in the fight over persistent pesticides, which has been much less one-sided than most people suppose. Inertia, strong economic interests with powerful Congressional connections, and the machinery of litigation and regulation have slowed the environmental charge against these compounds. Even before DDT was done, the environmentalists were working against the other persistent pesticides. By the end of 1970 the EDF had petitioned HEW for zero tolerance levels on dieldrin and aldrin (on the grounds they were carcinogenic), filed suits against heptachlor and chlordane, and was in federal court seeking to stop the USDA's plans to use Mirex against the fire ant (a revival of the 1957 campaign, which had used heptachlor). Two years later the EDF was working against almost all the persistent chlorinated pesticides on the market, seeking both cancellation and suspension. HEW refused to suspend use of the chemicals as human health hazards and EDF and EPA became involved (sometimes as allies, sometimes as adversaries) in a long series of actions. As each specific use had to be challenged and the regulatory process was long and complicated (the dieldrin–aldrin hearings took thirteen months of testimony), results have come slowly.

In June 1972, EPA ended most uses of dieldrin, but not until October 1974 did EPA administrator Russell Train announce a ban on the manufacture of dieldrin and aldrin (and Shell Chemical Company promptly appealed that decision). Finally, in May 1975, the District of Columbia Court of Appeals upheld the EPA, and Shell announced it would end production. In between there was a series of decisions, including one that the EDF appealed to the federal district court in Washington on the grounds that the administrator should have stopped all use, not just the production of dieldrin, and Shell appealed to the fifth circuit court, in New Orleans, asking that the decision to stop production be overturned. In late 1974, the EPA cancelled most uses of heptachlor and chlordane (though it again denied the need for suspension). Again, the proceedings dragged on through hearings, appeals, rehearings, and other actions, and ended with a suspension order in 1975. Mirex, the chemical planned against the fire ant, has been a particularly resistant chemical

with as many lives as the proverbial cat. In May 1979, the
House Agriculture Committee voted to allow a one-year
emergency use of Mirex against the fire ant, attaching the
provision to a bill to extend FIFRA for another year, a
provision rejected by the whole House in November. Other
cases have gone the same route, and as of mid-1979 only six of
the nineteen pesticides identified by the Mrak Commission as
possible human health hazards have been banned, a record the
EDF considers grossly inadequate.[11]

The campaign against persistent pesticides has changed and
developed since the first DDT suits, becoming a large,
carefully organized, and scientifically backed crusade against
environmental damage and environmental hazards. The EDF
started work against PCBs in 1970 and went on from there to
attack other compounds in the environment, gradually pull-
ing together the scattered suits and concentrating on the
problem of human health. This subject had been an issue as
early as the Madison hearing, and with the National Cancer
Institute's report in 1969, the EDF began to emphasize the
hazards of carcinogenesis. The first petitions on dieldrin and
aldrin had stressed that problem, and the same argument,
with complementary evidence, was used in suits against
heptachlor, chlordane, and endrin. In January 1975, the EDF
newsletter, in a column headed "Cancer: An Environmental
Disease" announced a "Toxic Chemicals Program" that
would seek "by public education and legal action to reduce
the quantities of man-made toxic and carcinogenic chemicals
in the environment." The campaign that had begun seven
years ago "to protect environmental quality with an emphasis
on wildlife, had evolved into a campaign for cancer preven-
tion." Human health, it went on, is clearly linked to environmental quality, and the warning of a decade ago—that
wildlife damage was only the warning of more serious
problems—was now confirmed.[12]

The EDF has also changed. It has branched out, establishing offices in strategic locations throughout the country,
recruited many more scientists, and taken on many more
projects. It has changed its tactics. Litigation has become a less

important weapon in the EDF's arsenal, and even when it goes to court now it is not usually seeking new environmental rights, but the enforcement of established environmental rules. The maneuvering over persistent pesticides, for example, has proceeded through EDF pressure on EPA and HEW to enforce laws that are on the books, and where those agencies have opposed EDF, it has been on the ground that the law is inapplicable (that, for instance, dieldrin and aldrin were not imminent cancer hazards to the population). In addition EDF has turned to public education. In May 1979, it published a book, *Malignant Neglect*, on the hazards of environmental carcinogens and the lack of federal action against this menace. [13]

The changes in tactics reflect changes in public opinion and the legal situation, and the limitations of litigation as a tool. The experience with DDT showed that litigation was in some ways a clumsy method. It required a great deal of money, time, and legal and scientific talent; there were serious difficulties in the way of getting a hearing—even with favorable court decisions on standing and jurisdiction; and the complexity of the issues and the testimony taxed the abilities of even a well-prepared examiner. The waning of public enthusiasm for environmental causes and the proliferation of organizations competing for funds in this area have only increased the difficulties. Even before the DDT fight ended, the EDF had to cope with declining enthusiasm among its supporters. Many people came to feel that the fight was over, that the initial victories meant the battle was won. When it came to dealing with obscure pesticides that most people had never heard of and could hardly pronounce, it was even more difficult to sustain interest. The EDF also lost some of its news value. In 1968 a group suing for environmental rights was news; by 1972 it was almost commonplace. In addition, the hearings were dull and their length made them even duller. Even the most sensational murder trial, spread over thirteen months and featuring a parade of 250 witnesses (as the dieldrin-aldrin hearings did) would cease to occupy the public mind. There were also a host of other environmental concerns by 1972—the Alaska pipeline, the SST, the whales, to name a few—that diffused the attention of even the most ardent environmentalist. Finally, the discovery of other carcinogens in the environment blunted public anxiety. There was a widespread feeling—wrong but widespread—that anything would cause cancer.

Another factor changed both the value and scope of litigation; the passage of environmental laws and the establishment of agencies and regulations meant that environmental concerns were being built into the system. Companies could no longer build power plants or major industrial complexes without taking the environment into account. The Corps of Engineers—indeed all federal agencies—had to prepare environmental impact statements. Even major public initiatives, such as President Carter's energy plan, had to deal with environmental questions. In this context the EDF and other groups found it less profitable to sue than to work through public education and to lobby to change policies before they took effect.

The place of litigation in the environmental movement, though, is not the main focus of this piece. What does our experience with DDT say about the effect of the social context on scientists' work and on the use of science in making public policy? A brief review of the principal argument is in order. This work contends that the widespread use of pesticides, which paved the way for DDT, was a result of the social and economic situation of the American farmers, the nature of government and private support for scientific research on agricultural problems, and the private profits to be made from chemical insecticides fully as much as it was due to the effectiveness of chemicals in controlling insect pests. Economic entomologists came to favor chemicals because they offered a means of insect control that, because it gave dramatic, quick results, helped establish the value of the profession to the farmer. Regulation of pesticides—also established before DDT appeared—was conditioned by physicians' concepts of disease and health (which led them to pass over chronic and ill-defined conditions), but also by the location of the regulatory agency in the department, the USDA, whose major political supporters were hostile to low tolerance levels and regulation based on consideration of chronic illness. These policies set the parameters for research into the effects of DDT and for regulation of the chemical. The discovery of DDT's unexpected effects in the environment was the work of scientists or agencies not previously involved in pesticide regulation, and it came largely because the evidence of environmental degradation grew too startling and puzzling to ignore. In the same way the change in pesticide regulation was the result of actions by groups and individuals outside the normal network of regulation who worked out new methods of influencing policy.

The changed social context—the new climate of public opinion, the higher values placed on wildlife and the environment, and the new legislation and regulation that followed from that—also affected the direction of scientific research and brought scientists to consider new problems and new solutions. Of all the disciplines involved, economic entomology was the most affected. It is too much to say that experience with DDT brought a disenchantment with prevailing methods of control or reoriented the profession, but it did bring a renewed interest in, and support for, nonchemical methods of control, and checked the enthusiasm for chemicals that had dominated the profession for at least a generation. The high road to chemical control was closed, not only because of the problems that DDT caused in various agricultural ecosystems, but because the furor over the effects on the environment had given the public and the environmentalists an interest and a voice in the policy by which chemicals were approved and regulated. At the same time there was more support for the long-term biological investigations and long-term research projects necessary for biological controls. For some, the effects of the DDT experience went deeper. They began to question the loose alliance among the USDA, the agricultural experiment stations and state universities where they worked, and the chemical companies that funded so much of the research. Although they did not see their situation in theoretical terms, they began to examine the links between the conditions under which they worked and the research they were encouraged to carry on.[14]

Change came slowly, and among economic entomologists, began independently of *Silent Spring*. Shortly after DDT was introduced, entomologists working with functioning systems of natural control found drawbacks to DDT. It killed beneficial as well as harmful insects and upset natural balances, loosing old pests, which had been controlled by imported predators, making pests out of insects that had never caused trouble. During the next decade the creation of new pests, particularly in the heavily mechanized cotton fields of California, became a more serious problem. These entomologists watched with dismay as deadlier and deadlier materials were introduced to combat the resurgent insects. By the early 1960s, even as *Silent Spring* appeared, a few entomologists were seeking better ways to handle the problems, and by the late 1960s, research on new methods was well established.

Entomologists took several different lines in approaching the problems of insect control without broad-spectrum pesticides. Some concentrated on more specific insecticides that would degrade rapidly once their job was done—an attempt to do away with the disadvantages of chemicals without changing the nature of their work. Others began to investigate new, and sometimes exotic, methods of control—juvenile hormones, sex attractants, or the sterile-male technique—and to push them to their limits. Still others began to build on the foundation laid by biological control specialists early in the century, adding other techniques as needed. The last approach has grown into integrated pest management (IPM), which is now becoming a serious alternative to chemical controls.[15] It differs from the usual chemical controls not only in seeing only a small place for chemicals and in using a variety of control measures but, more fundamentally, in its approach. IPM concentrates on two factors generally given short shrift in chemical control—the working of the agricultural ecosystem and the population density of the insects that could damage the crop. The first factor is a recognition that only a minute proportion of the insects present are pests; the remainder are controlled by environmental variables, including predators. Control, then, aims at suppressing damaging insects without changing other parts of the ecosystem. The other assumption is a major change from earlier practice, based on an equally revolutionary definition of the entomologists' problem. Pests, traditionally, have been identified by species—the boll weevil, for instance, was a pest—and control measures were prescribed when the pest was present. Under IPM, the term "pest" refers, not to the species, but to a population large enough to do damage to the crop. The focus of the entomologists' attention is not on the species but on the interaction of the population with the crop—with the agricultural ecosystem, if you will. The definition is not static but dynamic, and the aim is not to kill the pests but to prevent damage. The distinction may seem subtle—the chemical control specialist, after all, would also claim to be preventing damage to crops—but it is real and has affected the entomologists' work. IPM moves economic entomologists back toward becoming field ecologists, not simply insect physiologists.

Nonchemical methods, though, did not necessarily lead to a change of heart about man and nature; the use of new materials was quite compatible with a continued emphasis on man over nature and even with the crusade to eradicate pests. "Total Population Management" is the name given by historian John Perkins to another school of thought that developed in response to the new situation, using new techniques for old ends. It found its initial impetus in the campaign conducted by Edward Knipling of the USDA to wipe out the screw-worm fly by releasing enormous numbers of sterile males. Knipling's idea—that one could eradicate a pest by overwhelming the normal males with sterile competitors who would produce no young—was a bold one. From the successful work against the screw-worm, Knipling and others went on to suggest eradication of several major pests, including the boll weevil, using the sterile–male technique, traps, sex attractants, and chemical sprays. By the mid-1970s they had managed to get a pilot project going in the South to test their ideas, and we may yet see the first eradication of a major and widespread imported pest from the United States.[16]

In the late 1960s, as broad-spectrum pesticides came under increasing attack, some economic entomologists also began to question the social context in which they worked. Robert van den Bosch was the most outspoken critic of the connections between government research, state academic institutions, and commercial chemical manufacturers. In *The Pesticide Conspiracy* (1978) he accused the administrations of the land-grant schools and the USDA of working with powerful politicians and the agri-chemical industry to push a particular brand of insect control—the companies' own. Agricultural research into economic entomology, van den Bosch charged, was the creature of the companies who funded research, companies that also controlled legislation and administration of pesticide regulation. He was not alone. Others, more circumspect, had made similar charges in the early 1950s. Unfortunately van den Bosch died just after his book appeared, and it remains to be seen what effect it will have on economic entomologists.[17]

Economic entomology was the only discipline that underwent so drastic a change as a result of DDT. Other branches of biology, particularly wildlife management, added new tools and concerns as a result of the episode and became much more aware of the influence of chemicals introduced into the environment, but they did not undergo a professional revolution. DDT did not, for them, strike at the heart of their ideas of their work and reputation, and hence did not induce any serious examination of their concept of their professional social responsibility. For those scientists who became involved, DDT changed the course of their research and even the course of their careers, but for biologists in general it remained an episode. There has been some change in scientists' ideas about their social responsibility. They are less reluctant to become involved in public issues, and more concerned about the use of their research results, but these changes are hardly due to DDT.

It should be clear that the terms in which the initial inquiry were phrased—how DDT affected scientists and science—were in part misleading. To speak of a scientific answer to the question of the problem of DDT, to ask for the opinion of "scientists," or to expect "scientists" to solve the "problem" is useless; the "problem" of DDT is not a scientific one. The point would not be worth making except that the opposite assumptions have dominated the public debate. Expert commissions were appointed to give the scientists' answer to the question whether or not DDT should be used; advocates appealed to "science"; and the public seems to have assumed that DDT was used because it was the scientific answer to insect control problems. DDT, though, came into use because of economic considerations and continued in use because it was well suited to a particular kind of agricultural economy. The argument against it was, in its essentials, that DDT was ultimately causing more harm (in a very long-term sense) than good, that it was dangerous to health, and that the other damage it caused was not worth the immediate economic gain. But the problems presented by DDT use were questions of value, and the struggle over banning it was over what values should be explicit (and implicit) in public policy. For this kind of problem there is no scientific answer. There is scientific information upon which to base a decision; scientists may, and should, tell the public what the consequences of a decision will or might be. The public, though, cannot turn to any group for an answer; it must itself make the decision and take the responsibility.

Analytical Terms

CONCENTRATIONS of pesticide residues in food and in tissue are normally expressed in parts per million (ppm), occasionally parts per billion (ppb). In experiments in which materials are fed to, or injected into, animals, the dosages are usually given as milligrams per kilogram of body weight (mg./kg.); one milligram per kilogram is equal to 1,000 parts per million. In the English system of measurement, the units are grains per pound (gr./lb.), where a grain is 1/7,000 of an avoirdupois pound. Conversion figures are

$$1 \text{ ppm} = 0.007 \text{ gr./lb.}$$
$$1 \text{ gr./lb.} = 143 \text{ ppm}$$

Use of English measure, though, is confined to English-speaking countries before World War II; scientific measures are now invariably in metric units.

APPENDIX B

DDT

DDT "DDT" is an abbreviation derived from the common name d(ichloro)d(iphenyl)t(richloroethane). The proper name for the insecticidal compound is 1,1,1 trichloro-2,2*bis* (parachlorophenyl)ethane. This compound, also called p,p′ DDT, is the major constituent of "technical DDT," which is a mixture of isomers. The most common contaminant of the technical product, the insecticide, is o,p′ DDT, which appears in the commercial product to the extent of 8 to 21 percent.[1] The structural formulas for these compounds are:

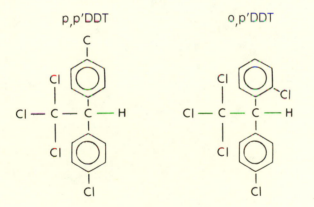

NOTE: ⬡ = benzene ring o = ortho p = para. Positions next to point of attachment of ring to body of molecule are ortho; opposite position is para.

DDE This is also derived from the common name of the compound, d(ichloro)d(iphenyl)e(thane). The proper name is

247

1,1 dichloro–2,2*bis* (parachlorophenyl)ethylene, its structural formula is;

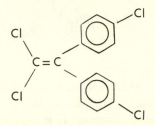

DDD or *TDE*   These are two abbreviations for the same compound, 1,1 dichloro–2,2*bis* (parachlorophenyl)ethane, another degradation production of DDT, which has also been used as an insecticide.

DDT in Human Fat

THE STORAGE OF DDT in human fat—and in the fat of other animals—is a consequence of its differential solubility. DDT is almost completely insoluble in water but much more soluble in nonpolar solvents including lipid tissue. Aquatic organisms thus concentrate DDT from the water, and organisms higher on the food chain concentrate the chemical further, absorbing the load in their prey. The level in human fat is a function of the concentration in the diet (the intake) and the metabolism of the compound in the body with subsequent excretion (the output). Both storage and the concentration of DDT in milk were known as early as 1944, and, with the widespread use of DDT in agriculture in the late 1940s, scientists began investigations of the extent of DDT storage in human fat. In 1950 Edwin P. Laug and his coworkers in the FDA's Division of Pharmacology undertook an analysis of fat samples from the general population in order to estimate accurately the level of DDT in the normal diet. Seventy-five fat samples from persons not occupationally exposed to DDT revealed an average level of 5.3 ppm DDT. Extrapolating from the results of rat tests, Laug estimated that the American diet contained about 0.05 ppm DDT.[1]

There was only brief resistance to this conclusion. Testifying before the Delaney Committee in 1951, Wayland Hayes suggested that the contamination of fat was due to accidental contact with DDT in agricultural areas, not to dietary contamination.[2] Public Health Service studies, however, did not support his conclusions. In January 1952, Leonard Scheele, surgeon general, submitted to the Delaney Committee a summary of the analyses done by Hayes's Technical Development Branch of the Public Health Service's Communicable Disease Center. One hundred eleven samples of fat from nonoccupationally exposed persons showed an average of 6.41 ppm DDT; samples from Wenatchee, Washington, a town located in a large apple-growing area, were not significantly different from those gathered in Savannah, Georgia (5.47 + 1.03 ppm against 6.62 + 1.75 ppm).[3]

Further studies confirmed the universal contamination but did not show significant increases in the concentration. Hayes's studies in 1956, on 51 convicts, and in 1958, on 227 people in the general population, showed about the same level. The results were in good agreement with Laug's with the exception of the extremes, which were missing in both of Hayes's studies.[4] Americans still have an approximate body burden of 15 to 20 ppm of DDT and DDT-related materials in their fat, though there seems to be some decline in the last five years.[5]

Enforcement of the Food
and Drug Laws

CONGRESS placed responsibility for enforcement of the Food
and Drug Act of 1906 in the Bureau of Chemistry, U.S.
Department of Agriculture, primarily because the head of that
bureau, Harvey W. Wiley, was the acknowledged leader of
the fight for the law. Despite Wiley's assertions that the law
gave him independent authority, the Secretary of Agriculture
soon established his claim to control his subordinate, beginning the ambiguous relation of the Secretary as both protector
of the farmer and the policeman of his produce. For Wiley's
account of the early years of enforcement (1906–1911), see his
History of a Crime Against the Food Law.[1] For a more balanced
account, read Oscar Anderson's *Health of a Nation*.[2]

Carl Alsberg, head of the bureau from 1912 to 1921,
reorganized the enforcement division and created the district
offices. Upon Alsberg's resignation in 1921, Walter G.
Campbell became acting head of the bureau. He had been,
successively, Wiley's chief inspector, first head of the eastern
district (when the district plan went into effect in 1914), and
Alsberg's assistant in charge of food and drug regulation. In
1923, when Charles Browne became chief of the Bureau of
Chemistry, Campbell became Director of Regulatory Work
in the Department of Agriculture, a new post created to
handle the department's regulatory responsibilities. Paul
Dunbar, another of "Wiley's boys" became his assistant.

In 1927 the position of Director of Regulatory Work was
abolished and the Food, Drug, and Insecticide Administration
was formed to administer all the law enforcement of the
department. In 1930 the enforcement of the Insecticide Act of
1910 was taken from this bureau and the name changed to
Food and Drug Administration. In 1940 the Food and Drug

Administration was transferred from the Department of Agriculture to the Federal Security Administration, the agency that became the cabinet-level Department of Health, Education, and Welfare in 1953. Campbell retired in 1944; his successor, Paul B. Dunbar, left in 1952—the last of the Wiley appointees to head the agency.

Regulatory work is described in (successively) the *Report of the Chemist* (1907–1927), *Report of the United States Food, Drug, and Insecticide Administration* (1928–1930), *Report of the Chief of the Food and Drug Administration* (1931–1940) (These documents are filed in the records of the Department of Agriculture.), *Annual Report of the Federal Security Administrator* [1940–1953]; Federal Security Administration), and annual reports of the Department of Health, Education, and Welfare (1954–1979).

APPENDIX E

Production of Pesticides

Some idea of the impact of DDT on the insecticide industry and on farming may be gained by comparing the production of the various insecticides. The first table shows annual production and dollar value of the three most common insecticides during the 1930s, the last decade before the use of DDT. Calcium arsenate was then the major insecticide for cotton insects, lead arsenate for fruit pests, and pyrethrum for household insecticides and vegetable dusts.

TABLE E-1
Production of Pesticides during the 1930s

	1931	1935	1937
calcium arsenate			
pounds	26,128,620	43,295,354	37,001,959
dollars	$1,279,789	$2,322,394	$1,879,253
lead arsenate			
pounds	37,974,038	52,145,851	63,291,440
dollars	$3,674,422	$4,173,462	$5,540,885
pyrethrum			
pounds	4,739,449	4,384,922	7,100,682
dollars	$1,850,647	$1,408,511	$2,021,751

SOURCE: U. S. Department of Commerce, Bureau of the Census, *Biennial Census of Manufactures, Part I* (Washington: Government Printing Office, 1939), 711.

Production figures in pounds from after the war show the changes, and the extent, of DDT use.

253

TABLE E-2
Production of Pesticides after World War II

	1953	1957	1959
DDT	37,904,000	83,981,000	124,545,000
lead arsenate	16,866,000	14,192,000	11,920,000
calcium arsenate	16,006,000	7,260,000	19,478,000

SOURCE: U. S. Department of Agriculture, Commodity Stabilization Service, *The Pesticide Situation* (Washington: Government Printing Office, annual, 1953–1960).

The price of DDT quickly fell from over a dollar a pound in 1945 to about twenty-five cents a pound by the mid-1950s. This was about the cost of lead arsenate, but since control with DDT required less pesticide and fewer applications, the latter compound was far less expensive to the farmer.

Unlike the arsenates and the botanicals, which had built up their markets and production facilities slowly, DDT entered the civilian agricultural market as a bulk chemical manufactured by large corporations. The firms under wartime contracts included Monsanto, Merck, Dupont, and Hercules.[1] These, and other large companies, continued to dominate the DDT market as primary manufacturers. In 1969 the manufacturers of DDT were Montrose Chemical Company of California, Diamond Shamrock, Olin Mathieson, Allied Chemical and Lebanon Chemical.

Tolerance Levels for
Pesticide Residues on Food

THE FOLLOWING TABLE shows the changes in arsenic and lead tolerance levels on fresh fruit during the period before World War II.

TABLE F-1
Tolerance Levels for Arsenic and Lead

Date	gr./lb	Date	gr./lb
ARSENIC			
1927	0.025	1930	0.015
1928	0.020	1931	0.012
1929	0.017	1932	0.010
LEAD			
1933 (21 Feb.)	0.025	1936	0.018
1933 (2 Apr.)	0.014	1937	0.018
1933 (20 Jun.)	0.020	1938	0.025
1934	0.019	1939	0.025
1935	0.018	1940	0.050

Arsenic tolerance levels are expressed, not as the weight of the element present, but as the weight of As_2O_3 formed, since this was assumed to be the most poisonous form. Although there was no explicit lead tolerance level before 1933, government workers assumed that the recommended washes would remove both lead and arsenic; the tolerance level on the latter in that case meant an implicit tolerance level on the former.

In 1945 the Food and Drug Administration set an interim level for DDT on most foods at 7 parts per million. Because of the properties of DDT, though, the controversy over the chemical never revolved around the accumulation from over-sprayed fruit, and the actual levels never became a subject for public debate.

Vapor Phase Chromatography

UNTIL THE WIDE-SPREAD USE of vapor phase chromotography in the early 1960s, the preferred method of analysis for DDT and its decomposition products was some variant of the Schecter-Haller method, developed by the BEPQ during World War II. This method depended on extraction of DDT "intensive nitration to polynitro derivatives and the production of intense colors upon addition of methanolic sodium methylate to a benzene solution of the nitration products."[1] The amount of DDT or related compounds could then be measured colorimetrically; p,p' DDT and p,p' DDT gave characteristic blue colors, and o,p' DDT gave violet-red color. DDE and DDA gave red colors.[2] Most early investigators used filters to isolate the DDT color and did not report the derivatives.

Gas chromatography was a major improvement on this method, under suitable conditions it was more sensitive, and it reported the presence and amount of several components rather than just one. The method depends on the reversible reaction of materials with a substrate. The sample, dissolved in an inert liquid, is injected into a tube filled with a material capable of retarding the flow of the components of the sample. A carrier gas sweeps the material down the tube, while the substrate, by retarding the various chemical components, separates them. Each chemical, then, emerges from the apparatus and enters the detector at a specific time dependent on the material, the speed of the gas, and the temperature. Confusion arises when two chemicals enter the detector at once. Common cross-checks to eliminate this interference including using another method of analysis and running identical samples under different conditions.

NOTES

INTRODUCTION

1. Clay Lyle, "Achievements and Possibilities in Pest Eradication," 5–6. "DDT Considered Safe for Insecticidal Use." *Chemical and Engineering News* 23 (10 January 1945), 61, and 23 (25 September 1945), 1632.

2. Rachel Carson, *Silent Spring*, 22.

3. Lorrie Otto, one of the people who helped bring the EDF to Wisconsin, mentioned (author's interview, 18 February 1974) that she had not even read all of *Silent Spring*, it was too familiar and discouraging. None of the members of the EDF mentioned *Silent Spring* as an influence on their thinking. Woodwell, in fact, had corresponded with Carson when she was writing the book, but said that he was "sufficiently suspicious of her approach to write her with extraordinary restrictions on the use of my information." (letter to author, 4 April 1977). In the same letter he strongly defended the book as a "classic."

CHAPTER 1

1. E. G. Packard, *Entomological News* 17, 256 (1906), cited in James C. Whorton, *Before Silent Spring*, 91.

2. On the rapid development of DDT manufacturing, see "Industrial News," *Chemical and Engineering News* 22 (25 February 1944), 44, and 22 (10 July 1944), 1113. *New York Times*, 6 December 1944, 31:3; 29 December 1944. 26:5. On the release of DDT, see *Chemical and Engineering News* 23 (25 August 1954), 1442. A review of DDT work to 1945 is contained in Stanley J. Cristol and H. L. Haller, "The Chemistry of DDT—A Review," *Chemical and Engineering News* 23 (25 November 1945), 2070–2075.

3. Whorton, *Before Silent Spring*, chap. 1.

4. The problems of insect infestation are most clearly revealed in the annual "Report of the Entomologist." See also Whorton, *Before Silent Spring*, chap. 1.

5. Leland O. Howard, "Danger of Importing Insect Pests," 529–552.

6. For a short account of the rise of chemical controls in economic entomology, see Thomas R. Dunlap, "The Triumph of Chemical Pesticides in Insect Control, 1890–1920."

7. U. S. Department of Agriculture, *Annual Report of the United States Department of Agriculture, 1868.* 438. Leland O. Howard, "Progress in Economic Entomology in the United States," 137.

8. The rise of insecticides may be traced through the annual "Report of the Entomologist" (full citation in note 4). See also Dunlap, "Triumph of Pesticides" and Edward O. Essig, *A History of Entomology,* (New York: Macmillan, 1931), chap. 7.

9. Leland O. Howard, *A History of Applied Entomology,* 78–81.

10. Ibid., 82.

11. Ibid., 75.

12. Anderson Hunter Dupree, *Science in the Federal Government,* 158–162. On the relations between the government and the stations, see Charles E. Rosenberg, "Science, Technology and Economic Growth: The Case of the Agricultural Experiment Station Scientist, 1875–1914," 1–20 and "The Adam Act: Politics and the Cause of Scientific Research," 3–12.

13. Dupree, *Science,* 162. Leland O. Howard to Stephen A. Forbes, 20 March 1911, in General Correspondence, 1908–1924, Record Group 7, "Records of the Bureau of Entomology and Plant Quarantine," National Archives.

14. Howard, *Applied Entomology,* 189 and "The Organization Meeting of the Association of Economic Entomologists at Toronto, August 1889," 26–30. Reminiscences by various colleagues of his generation follow, 30–44.

15. Dupree, *Science,* 158–162.

16. On the development of pesticide regulation, see Adelynne Hiller Whitaker, "A History of Federal Pesticide Regulation in the United States to 1947."

17. Dupree, *Science,* 159.

18. Stephen A. Forbes, "The Ecological Foundations of Applied Entomology," 7. A. G. Ruggles, "Pioneering in Economic Entomology," 39.

19. William Moore, "The Need of Chemistry for the Student of Entomology," 172.

20. Letter cited in Howard, *Applied Entomology,* 125.

21. C. H. Tyler Townsend, "Report on the Mexican Cotton Square-and-Boll Weevil (*Anthonomus Grandis* Boheman) in Texas," 17. On boll weevil work in the South, see Douglas Helms, "Just

Lookin' for a Home: The Cotton Boll Weevil and the South" (Ph.D. dissertation, Florida State University, 1977).

22. Howard, *Applied Entomology,* 127. Walter D. Hunter, "The Control of the Boll Weevil" and "Methods of Controlling the Boll Weevil." Mason Snowden, "Cotton Growing in Louisiana: Demonstration Methods Under Boll Weevil Conditions," 18.

23. C. H. Tyler Townsend to Leland O. Howard, 3 December 1895 in Townsend Correspondence.

24. Charles L. Marlatt, "Boll Weevil Notes, April 24 to May 6, 1896," copy in Townsend Correspondence.

25. *Houston (Texas) Post,* 15 December 1904, in Newspaper Clippings Relating to the Boll Weevil, 1901–1912, Southern Field Crop Insect Investigation (SFCII). R. E. McDonald, "The Boll Weevil: A Review of the Methods of Control," 9. See also clippings in Newspaper Clippings, 1904, SFCII.

26. Walter D. Hunter, "The Use of Paris Green in Controlling the Cotton Boll Weevil," 29–32.

27. Hunter, "The Control of the Boll Weevil, 5.

28. On bureau attempts to work through bankers and landlords, see Marlatt, "Boll Weevil Notes," 10. On crop rotation, see Mason Snowden, "Cotton Growing in Louisiana: Demonstration Methods under Boll Weevil Conditions," and Howard, *Applied Entomology,* 125–132.

29. Fred Malley, State Entomologist of Texas. Undated clipping, 1902, Newspaper Clipping, SFCII. McDonald, "Boll Weevil."

30. Snowden, "Cotton Growing."

31. Hunter, "Control of the Boll Weevil," 24–25.

32. Townsend to Howard, 3 January 1898 in Townsend Correspondence.

33. Howard to Townsend, 7 June 1898, "Letters Sent, 1893–1908," Record Group 7, National Archives.

34. David F. Houston, "Study and Investigation of Boll Weevil and Hog Chlorea Plagues." Patent 954,629, copy from Special Correspondence Files, Records of the Division of Cotton Insect Investigation, Record Group 7, National Archives.

35. Two drawers of newspaper clippings on the Guatemalan Ant are in Newspaper Clippings, SFCII. Walter D. Hunter to Howard, 12 July 1904, 7 March 1905, 4 April 1905, and Expense Sheet in Special Correspondence, Division of Cotton Insects, Record Group 7, National Archives.

36. Townsend to Howard, 24 October 1894, 18 July 1895, 19 November 1895 in Townsend Correspondence.

37. Walter D. Hunter and W. Dwight Pierce, "The Mexican Cotton Boll Weevil: A Summary of the Investigation of this Insect up to December 3, 1911," 143. W. Dwight Pierce, R. A. Cushman, and C. E. Hood, "The Insect Enemies of the Cotton Boll Weevil."

38. Hunter, "Use of Paris Green," 21–22. Howard, *Applied Entomology*, 131. B. R. Coad, "Killing Boll Weevils with Poison Dust," 241. "Report on the Accomplishments of the Work on Cotton Insects Investigations of the Bureau of Entomology and Plant Quarantine and Preceding Organizations of the United States Department of Agriculture from 1854 to 1938," 14–15, 17. B. R. Coad, "Recent Experimental Work on Poisoning Cotton-Boll Weevils," 2.

39. Coad, "Killing Boll Weevils," 241.

40. Warren E. Hinds, "Boll Weevil Control for Louisiana, 1924," 4; "Killing Over-Wintering Weevils: the Pre-Square Poisoning"; "Machines and Poison for Weevil Control"; "Poisoning the Boll Weevil on Fruiting Cotton: The Regular Dusting Treatment."

41. Walter D. Hunter and B. R. Coad, "The Boll Weevil Problem," 14.

42. Coad, "Dusting Cotton from Airplanes." Eldon W. Downs and George F. Lemmer, "Origins of Aerial Crop Dusting," 130–132.

43. "Report of the Entomologist, 1930,"12.

44. Howard, *Applied Entomology*, 500–503. Leland O. Howard, "The Practical Use of Insect Enemies of Injurious Insects," 273–278.

45. On inflated expectations, see Howard, *Applied Entomology*, 503, and Harry S. Smith, "What May We Expect from Biological Control," 506–511.

46. Massachusetts State Board of Agriculture, *Agriculture in Massachusetts*. For federal work, see "Report of the Entomologist" and Howard, *Applied Entomology*, 115–119.

47. *Medford [Massachusetts] Mercury, Ravages of the Gypsy Moth in Medford;* 33 Stat 1269, "The Gypsy and Brown-Tail Moths," 18–19. Clipping from *Boston Transcript,* 11 July 1907. B. F. Thorpe, ed., *[Bangor] Maine Farmer,* to Leland O. Howard; Howard to Thorpe, 11 September 1905. All material in Correspondence Relating to Investigations of the Gypsy and Brown-Tail Moths, 1905–1908, Record Group 7, National Archives. This collection hereinafter cited as Gypsy Moth, Record Group 7.

48. W. F. Fiske to Leland O. Howard, 23 April 1908, reply from Howard, 25 April 1908; Howard to William D. Soplier, 6 May 1908; Howard to A. E. Stone, superintendent of Rhode Island Gypsy

Moth Suppression, 6 January 1908. All from Gypsy Moth, Record Group 7.

49. Leland O. Howard and W. F. Fiske, "The Importation into the United States of the Parasites of the Gypsy Moth and the Brown-Tail Moth," 306–310.

50. Correspondence between Albert F. Burgess and Leland O. Howard, in Correspondence of Albert F. Burgess, General Correspondence, Record Group 7, National Archives. Albert F. Burgess, "The Present Status of the Control of the Gypsy Moth and the Brown-Tail Moth by Means of Parasites."

51. Howard, *Applied Entomology*, 505.

52. Harry S. Smith played a leading role in the continuation of biological work in California and trained a generation of successors. See Arnold Mallis, *American Entomologists*, 501–503.

53. William Moore, "The Need of Chemistry," 172.

54. C. W. Woodworth, "The Insecticide Industry in California," 358. The same complaint may be found in contemporary literature: Robert van den Bosch, *The Pesticide Conspiracy*.

55. Paul DeBach, *Biological Control by Natural Enemies*, 86.

56. Wilton E. Britton, "The Academic Training of the Entomologists in College and Experiment Stations of the United States." Leland O. Howard, "The Education of the Entomologists in the Service of the U.S. Department of Agriculture." John Henry Comstock, "The Present Methods of Teaching Entomology." Paul Knight, "The Development and Present Status of Entomological Courses in American Colleges and Universities."

57. Ruggles, "Pioneering"

58. Leland O. Howard, *Fighting the Insects: The Story of an Entomologist* and *The Insect Menace*.

59. R. Keith Chapman, author's interview, 1 March 1974.

60. Harriet A. Geer, Randall Lotta, and Arthur W. Lindquist, "The Dispersal of Insecticides."

61. Clay Lyle, "Achievements and Possibilities in Pest Eradication," 6.

62. Testimony of Orville Freeman, Secretary of Agriculture in U.S. Congress, Senate, Subcommittee on Reorganization and International Organization of the Committee on Government Operations, *Interagency Coordination on Environmental Hazards (Pesticides)*, 88.

Chapter 2

1. Charles H. Fernald, "The Evolution of Economic Entomology," in *AAEE—Presidential Addresses, 1889–1911,* 7 (each address separately paged). Leland O. Howard, "Progress in Economic Entomology in the United States," 147. James C. Whorton, "Insecticide Residues on Foods as a Public Health Problem: 1865–1938," 31. The discussion of medical reaction to pesticides and early federal regulation in this chapter is based, to a large extent, on James Whorton's work, published as *Before Silent Spring.* I read Whorton's dissertation and *Before Silent Spring* when it was submitted to the publisher, and carried on a correspondence with Whorton for several years. In addition, his major advisor, Professor Aaron Ihde, was one of the members of my dissertation committee.

2. Anton J. Carlson to W. C. O'Kane, 20 November 1914, in W. C. O'Kane, C. H. Hadley, Jr., and W. A. Osgood, "Arsenical Residues after Spraying."

3. On physicians' relative unconcern with this problem, see Whorton, *Before Silent Spring,* chap. 2.

4. *New York Times,* 25 September 1891, 9; 26 September 1891, 2; 27 September 1891, 16.

5. Beverley T. Galloway, "The Grape Scare in New York," 375–376. *New York Times,* 29 September 1891, 4.

6. Harvey W. Wiley, *The History of a Crime against the Food Law.* For a more temperate account, see Oscar E. Anderson, *The Health of a Nation: Harvey W. Wiley and the Fight for Pure Food.*

7. For a summary of the changes in food-law enforcement, see Centennial Committee, *A Century of Service: The First 100 Years of the USDA,* 48–49, 474–475, 518.

8. J. C. Folger, "The Commercial Apple Industry in the United States," 367–377. Albert Hartzell and Frank Wilcoxon, "Analyses of Sprayed Apples for Lead and Arsenic." U. S. Department of the Interior, Geological Survey, *The National Atlas of the United States of America,* 97–99. Paul A. Neal, Waldemar C. Dressen, Thomas I. Edwards, Warren H. Reinhart, Steward H. Webster, Harold T. Caslenberg, and Lawrence T. Fairhall, "A Study of the Effect of Lead Arsenate Exposure on Orchardists and Consumers of Sprayed Fruit," 3. The remarks in this section apply to the apple and pear industry. These two fruits gave the bureau the most trouble, both because they were the most popular and because they were grown in semi-arid areas.

9. Whorton, *Before Silent Spring,* chap. 4.

10. An example of what the growers feared is the cranberry scare of 1959, when the announcement that the FDA had seized some cranberries contaminated with a chemical that caused cancer in mice caused a sharp drop in sales of cranberry sauce.

11. Walter S. Frisbee, "Federal Control of Spray Residues on Fruits and Vegetables," 370. H. E. Cox, "The Occurrence of Arsenic in Apples," 135.

12. Frisbee, "Federal Control of Spray Residues," 370.

13. Harvey W. Wiley, "The Poison on our Fruits."

14. U. S. Department of Agriculture, *Report of the Secretary of Agriculture for 1926* (Washington: Government Printing Office, 1927), 48.

15. Ruth deForest Lamb, *American Chamber of Horrors* 215. Lamb was a Department of Agriculture information officer and wrote her book to popularize the FDA-sponsored revision of the food and drug laws in the 1930s.

16. Wiley's account of the Remsen board is in his *History of a Crime Against the Food Law.* See also Anderson's *Health of a Nation* and Theodore Roosevelt's comments on Wiley in his letter to Henry Rusby, a drug expert in the Bureau of Chemistry, 7 January 1909 in Elting E. Morison, ed., *Letters of Theodore Roosevelt,* vol. 6, 1467–1468.

17. Lamb, *American Chamber of Horrors,* 218.

18. See Appendix A—Analytical Terms

19. Report of conference at the Bureau of Chemistry, 3 January 1927. Records of the Food and Drug Administration, General Spray Residue File, 1935. National Record Center, Record Group 88, cited in Whorton, *Before Silent Spring,* 186.

20. Lamb, *American Chambers of Horrors,* 228, 233.

21. See Appendix F—Tolerance Levels.

22. See Appendix F—Tolerance Levels.

23. Charles O. Jackson, *Food and Drug Legislation in the New Deal;* "Notice to Growers and Shippers of Fruit," 2 April 1933, in U. S. Congress, Senate, Subcommittee of the Committee of Commerce, *Foods, Drugs, and Cosmetics,* hearings before a Subcommittee of the Committee on Commerce on S.5, 451. (hereinafter cited as *Hearings on S.5*) For a full description of this episode, see Jackson, *Food and Drug Legislation.*

24. David F. Cavers, "The Food and Cosmetic Act of 1938: Its Legislative History and Its Substantive Provisions," 15. "Notice to Growers and Shippers of Fruit," in *Hearings on S.5,* 451–452.

25. See Appendix F—Tolerance Levels.

26. W. C. Geagley, "State Control of Spray Residues," 374–375.

27. Paul A. Dunbar, "Memories of Early Days of Federal Food and Drug Law Enforcement," 129.

28. T. Swann Harding, "How Much Poison Can We Eat?" Lamb, *American Chamber of Horrors*, 215–230.

29. U. S. Department of Agriculture, *Report of the Chief of the Food and Drug Administration for 1935* (Washington: Government Printing Office, 1936), 3.

30. C. N. Myers, L. Van Dych, and Binford Throne, *Medical Times* (May 1929); C. N. Myers and Binford Throne, *New York State Journal of Medicine* (15 October 1929), cited in C. N. Myers, Binford Throne, Florence Gustafson, and Jerome Kingsbury, "Significance and Danger of Spray Residue,"

31. Myers, "Significance of Spray Residues," 628.

32. Unsigned editorials in *Journal of the American Medical Association*, "Dangers of Lead Arsenate as Spray in Orchards," "Spray Residues on Foods," "Hazards of Contaminated Fruits and Vegetables."

33. Fred B. Linton, "Leaders in Food and Drug Law" (part 4 of 5 parts) 482.

34. Linton, "Leaders in Food and Drug Law," (4) 432–483.

35. Linton, "Leaders in Food and Drug Law," (4) 485.

36. Fraser testified at *Hearings on S.5*, 254–276, and at another hearing on this subject before the U. S. Congress, Senate, Committee on Commerce, *Foods, Drugs, and Cosmetics*, hearings before a Subcommittee of the Committee on Commerce, U. S. Senate, on S.2800, 417–451. (hereinafter cited as *Hearings on S.2800*)

37. *Hearings on S.2800*, 444.

38. Jackson, *Food and Drug Legislation*, 189.

39. Linton, "Leaders in Food and Drug Law" (4), 486–487.

40. Neal et al., "Lead Arsenate Exposure," 4.

41. Neal et al., "Lead Arsenate Exposure, ix. Emphasis is in original.

42. Federal Security Administration, *Report of the Chief of the FDA, 1940* (Washington: Government Printing Office, 1941), 297.

43. Cavers, "Food, Drug, and Cosmetic Act of 1938," 3.

CHAPTER 3

1. Hans Zinsser, *Rats, Lice, and History*; William H. McNeil, *Plagues and People*.

266 NOTES, PP.60–63

2. Othmar Zeidler, "Verbindung von Chloral mit Brom- und Chlorbenzol." P. Lauger, H. Martin, and P. Muller, "Uber Konstitution und toxische Werkung von natürlichen und neuen synthetischen insektentötenden Stoffen." John H. Perkins, "Reshaping Technology in Wartime: The Effect of Military Goals on Entomological Research and Insect-Control Practices."

3. Edward F.Knipling, "DDT Insecticides Developed for Use by the Armed Forces," 205; Richard Ormsbee, "The Toxicology and Mechanism of Action of DDT in Mammals," 622; and H. L. Haller and Ruth L. Busby, "The Chemistry of DDT," 618–619.

4. Harriet A. Geer, Randall Lotta, and Arthur W. Lindquist, "The Dispersal of Insecticides," 636–645; New York Times, 6 December 1944, 31:3. Wartime tests are described in Ormsbee, "Toxicology of DDT," and in Paul A. Neal, W. F. von Oettingen, W. W. Smith, R. B. Malmo, R. C. Dunn, H. E. Moran, T. R. Sweeney, D. W. Armstrong, and W. C. White, "Toxicity and Potential Dangers of Aerosols, Mists, and Dusting Powders containing DDT"; testimony of Paul A. Neal in U. S. Congress, House, House Select Committee to Investigate the Use of Chemicals in Food Products, Chemicals in Food Products; and Paul A. Dunbar, "The Food and Drug Administration Looks at Insecticides," 236.

5. W. A. Brittin, "Chemical Agents in Food," 594.

6. C. R. S. Harris, Allied Military Administration of Italy, 1943–1945, 419–420; Charles M. Wheeler, "Control of Typhus in Italy 1943–1944 by the Use of DDT"; Fred L. Soper, W. A. Davis, F. S. Markham, and L. A. Riehl, "Typhus Fever in Italy, 1943–1945, and its Control with Louse Powder"; H. L. Haller and Stanley J. Cristol, "The Development of New Insecticides."

7. "DDT Considered Safe for Insecticidal Use."

8. Harris, Allied Military Administration, 422; F. C. Bishopp, "Insect Problems in World War II with Special Reference to the Insecticide DDT," 374.

9. Geer, "Dispersal of Insecticides"; J. S. Yuill, D. A. Isler, and George D. Childress, "Research on Aerial Spraying."

10. Geer, "Dispersal of Insecticides"; Yuill et al., "Research on Aerial Spraying."

11. George C. Decker, "Agricultural Applications of DDT, with Special Reference to the Importance of Residues," 553; James C. Leary, William J. Fishbein, and Lawrence C. Salter, DDT and the Insect Problem, 69; Chemical and Engineering News 23 (25 August 1945), 1442.

12. Adelynne Hiller Whitaker, "A History of Federal Pesticide Regulation in the United States to 1947" 385–387.

13. On the enthusiasm for DDT, see O. T. Zimmerman and Irvin Levine, *DDT: Killer of Killers*.

14. The FDA's "provisional" tolerance levels simply meant that the agency would seize any crops with residues over this level and begin court proceedings to have them condemned as unfit for human consumption. A formal tolerance level, which could only be set after public hearings, effectively had the force of law and could not be challenged by the grower except through the ten circuit courts. For an analysis of this provision of the 1938 Food, Drug, and Cosmetic Bill, see Charles O. Jackson, *Food and Drug Legislation in the New Deal*, 189.

15. Federal Security Administration, Food and Drug Administration, *Annual Report of the FDA, 1946* (Washington: Government Printing Office, 1946), 6.

16. Paul A. Neal, W. F. Van Oettingen, R. C. Dunn, and N. E. Sharpless, "Toxicity and Potential Dangers of Aerosols and Residues from Such Aerosols Containing Three Percent DDT."

17. R. A. M. Case, "Toxic Effects of 2,2*bis* (p-chlorophenyl) 1,1,1 trichloroethane (DDT) in Man."

18. Ormsbee, "Toxicology of DDT," 634.

19. "Symposium—Toxicity of DDT." Decker later became convinced that DDT was safe and opposed further regulation of DDT.

20. American Medical Association, Council on Foods and Nutrition, "Health Hazards of Pesticides," *Journal of the American Medical Association* 137 (23 August 1948), 1603; see also "Pesticides: Chemical Contaminants of Foods."

21. Figures on pesticide production and use may be found in U. S. Department of Agriculture, *The Pesticide Situation* and its successor *The Pesticide Review* (Washington: Government Printing Office, annual). The most easily accessible account of the various uses of DDT is U. S. Department of Agriculture, *Insects, the Yearbook of Agriculture for 1952* (Washington: Government Printing Office, 1952). More professional accounts may be gleaned from the *Journal of Economic Entomology*.

22. *Annual Report of the FDA, 1949*, 11. Testimony of Fred C. Bishopp in U. S. Congress, House, Select Committee to Investigate the Use of Chemicals in Food Products, *Chemicals in Food Products*, 82nd Congress, 405, 530–540. (hearings hereinafter cited as *Chemicals in Food Products*)

23. Edwin P. Laug, Frieda M. Kunze, and C. S. Prickett, "Occurrence of DDT in Human Fat and Milk," 246. See Appendix C—DDT in Human Fat.

24. Brittin, "Chemical Agents in Food"; R. K. Phelan, "Problems in Manufacturing Baby Food."

25. Hearings held in 1950 and 1951, cited as *Chemicals in Food Products*, 81st and 82nd Congresses.

26. Testimony of Paul A. Dunbar in *Chemicals in Food Products*, 81st Congress, 31–35.

27. Dunbar in *Chemicals in Food Products*, 81st Congress, 236.

28. Testimony of Arnold Lehman in *Chemicals in Food Products*, 81st Congress, 338.

29. Testimony of John Dendy in *Chemicals in Food Products*, 82nd Congress, 220–223. Dendy's employer, the Texas Research Foundation, is a private foundation dedicated to agricultural research on Texas soils. Testimony of Robert A. Kehoe in *Chemicals in Food Products*, 81st Congress, 760; testimony of Wilhelm C. Heuper in *Chemicals in Food Products*, 82nd Congress, 1370.

30. Testimony of L. G. Cox in *Chemicals in Food Products*, 82nd Congress, 594.

31. Testimony of Lea S. Hitchner in *Chemicals in Food Products*, 81st Congress, 368.

32. *Chemicals in Food Products*, 81st Congress, 374.

33. *Chemicals in Food Products*, 82nd Congress, 181–185.

34. Testimony of Samuel Fraser in *Chemicals in Food Products*, 81st Congress, 688.

35. Testimony of C. E. Chase in *Chemicals in Food Products*, 82nd Congress, 625.

36. Testimony of Wayland J. Hayes, Jr., in *Chemicals in Food Products*, 82nd Congress, 95.

37. Testimony of Wayland J. Hayes, Jr., in *Chemicals in Food Products*, 82nd Congress, 101. See Appendix C—DDT in Human Fat.

38. It became law on 22 July 1954, 68 Stat. 511.

39. The best single source for industry comment is *Agricultural Chemicals*, which provided a forum for industry comment and criticism from the hearing through the passage of the bill.

40. W. W. Sutherland, "No More Legislation Needed," 43.

41. Frank Princi, "Toxicology and Hazard Record of the Newer Pesticides," 44.

42. "Chemicals in Food."

43. Edward F. Knipling, "The Greater Hazard—Insects or Insecticides."

44. D. A. Chant, professor and chairman of the Department of Zoology, University of Toronto, to author, personal communication, 15 November 1973. G. C. Ullyet, "Insects, Man, and the Environment."

45. Carl D. Fisher, "Pesticides, Past, Present and Future. Insecticides, Miticides, and Rodenticides."

46. *Agricultural Chemicals* assumed, to some degree, the job of reporting that had been done by *Oil, Paint, and Drug Reporter*. The splitting off of the more specialized publication shows that the business had become large enough to support a specialized publication. The articles and editorials illustrate the various concerns of the industry.

CHAPTER 4

1. Figures on pesticide production and use are in U. S. Department of Agriculture, Commodity Stabilization Service, *The Pesticide Situation* and its successor *The Pesticide Review* (Washington: Government Printing Office, annual).

2. Rachel Carson, *Silent Spring*. A sketch, somewhat biased, of the reaction to *Silent Spring* is in Frank Graham, Jr., *Since Silent Spring*.

3. Clarence Cottam and Elmer Higgins, "DDT and its Effect on Fish and Wildlife," 51.

4. Department of Lands and Forests, Ontario, Canada, *Forest Spraying and Some Effects of DDT;* Neil Hotchkiss and Richard H. Pough, "Effect on Forest Birds of DDT used for Gypsy Moth Control in Pennsylvania," 206.

5. Cottam and Higgins, "DDT and its Effect," 51.

6. L. W. Orr and L. O. Mott, "The Effects of DDT Administered Orally to Cows, Horses, and Sheep," 428–432. John H. Draize, Geoffrey Woodard, O. Garth Fitzhugh, Arthur A. Nelson, R. Blackwell Smith, Jr., and Herbert O. Calvery, "Summary of Toxicological Studies of the Insecticide DDT."

7. "Report of the Committee on the Relation of Entomology to Conservation," *Journal of Economic Entomology* 40 (February 1947), 149. Italics in original.

8. Clarence Cottam, speech before the National Parks Association, 2 October 1962, quoting a speech by George Decker made to the 29th Annual Conference of the North Central States Entomologists, 23–24 March 1950; copy from files Roland Clement, National Audubon Society.

9. Paul DeBach, "The Necessity for an Ecological Approach to Pest Control on Citrus in California," 445.

10. "Dutch Elm Disease Found Here," 3. U. S. Department of Agriculture, *Report of the Secretary of Agriculture: 1933* (Washington: Government Printing Office, 1933), 72.

11. Russell R. Whitten and Roger U. Swingle, "The Dutch Elm Disease and its Control." Roger U. Swingle, Russell R. Whitten and E. G. Brewer, "Dutch Elm Disease."

12. H. P. Blagbrough, "Reducing Wildlife Hazards in Dutch Elm Disease Control," *Journal of Forestry* 5, no. 6, 468–469, cited in George J. Wallace, Walter P. Hickell, and Richard F. Bernard, *Bird Mortality in the Dutch Elm Disease Program in Michigan*, 42.

13. Roy J. Barker, "Notes on Some Ecological Effects of DDT Sprayed on Elms," 271, 269–273.

14. Whitten and Swingle, "Dutch Elm Disease," 10. Not everyone accepted these arguments, and the environmentalists raised this issue in the late 1960s as part of their effort to discredit the use of DDT for Dutch elm disease control.

15. Wallace et al., *Bird Mortality*, 9.

16. Lorrie Otto, author's interview, 18 February 1974.

17. Ibid. Following paragraph is also based on Otto's account.

18. Wallace et al., *Bird Mortality*, 13.

19. Ibid.

20. George J. Wallace, "Insecticides and Birds." See also Wallace et al., *Bird Mortality*.

21. George J. Wallace in Wisconsin Department of Natural Resources, "In the Matter of the Petition of the CNRA of Wisconsin, Inc. et al. for a Declaratory Ruling on DDT," 2524.

22. Wallace et al., *Bird Mortality*, 4.

23. Joseph J. Hickey, author's interview, 16 July 1973.

24. Ibid.

25. L. Barrie Hunt, "Songbird Breeding Populations in DDT-Sprayed Dutch Elm Disease Communities," 140.

26. Joseph J. Hickey and L. Barrie Hunt, "Initial Songbird Mortality following a Dutch Elm Disease Control Program." Hickey interview.

27. Albert F. Burgess, "Gypsy Moth Barrier Zone Maintenance Problem" and "The Value to Uninfested States of Gypsy Moth Control Extermination."

28. Work on the gypsy moth may be traced in the series of articles by Burgess in the *Journal of Economic Entomology* and in the Correspondence of Albert F. Burgess, General Correspondence, 1908–1924, Records of the Bureau of Entomology and Plant Quarantine, Record Group 7, National Archives. A useful review, with comments on the period to 1960 is in Albert C. Worrell, "Pests, Pesticides, and People." See also W. L. Popham and David G. Hall, "Insect Eradication Programs," 250.

29. W. V. O'Dell, "The Gypsy Moth Control Program." Wilhelmine Kirby Waller, "Poison on the Land." *New York Times,* 16 May 1957, 33:3.

30. *New York Times,* 9 May 1957, 1:7; 25 May 1957, 4:3; 7 July 1957, 23:1.

31. *New York Times,* 11 February 1958, 33:7; 14 February 1958, 25:8.

32. *New York Times,* 22 February 1958, 6:4; 25 February 1958, 45:4.

33. *New York Times,* 24 June 1958, 33:5.

34. Robert L. Rudd, *Pesticides and the Living Landscape,* 34.

35. *New York Times,* 25 December 1957, 33:6; 27 December 1957, 33:6; 7 January 1958, 24:2.

36. Rudd, *Pesticides and the Living Landscape,* 34–36. *New York Times,* 23 December 1957, 25:8; 27 December 1957, 11:2.

37. George J. Wallace, "Insecticides and Birds."

38. Joseph J. Hickey, J. A. Keith, and Francis B. Coon, "An Exploration of Pesticides in a Lake Michigan Ecosystem."

39. Statement of John H. Baker to Subcommittee on Agricultural Appropriations, presented by Robert L. Burnap of the Audubon public relations staff, 24 March 1959, copy from papers of Roland Clement, Audubon Society.

40. Carl W. Buchheister, "Four of Six Conservation Goals Reached."

41. Collected press releases of Audubon Society on pesticides, files of Roland Clement.

42. Chandler S. Robbins, Paul F. Springer, and Clark G. Webster, "Effects of Five-Year DDT Application on Breeding Bird Population," 216.

43. James B. Dewitt, "Effects of Chlorinated Hydrocarbon Insec-

ticides upon Quail and Pheasants" and "Chronic Toxicity to Quail and Pheasants of Some Chlorinated Insecticides.

44. James B. Dewitt, "Effects of Chemical Sprays on Wildlife."

45. Barker, "Ecological Effects of DDT Sprayed on Elms," 269–274.

46. Rudd, *Pesticides and the Living Landscape,* 252.

47. Eldridge G. Hunt, "Pesticide Residues in Fish and Wildlife of California. Rudd, *Pesticides and the Living Landscape,* 250.

48. Ibid.

49. Hickey interview.

50. Charles L. Broley, "The Flight of the American Bald Eagle."

51. Alexander Sprunt IV, "An Eagle-eyed Look at Our Bald Eagle."

52. Alexander Sprunt IV, "Bald Eagles Aren't Producing Enough Young."

53. Hickey interview. Roland Clement, author's interview, 2 June 1977.

CHAPTER 5

1. Rachel Carson, *Silent Spring,* 23. "NACA Speakers Emphasize Industry's Role in Pesticide Safety"; University of Wisconsin Department of Entomology, "Critique of 'Insecticides and People,' " 16.

2. Carson, *Silent Spring,* 18.

3. Ibid, 22.

4. Ibid, 14.

5. Ibid, 168.

6. Rachel Carson, "A Report on Progress," in Paul Brooks, *The House of Life,* 244.

7. Joseph J. Hickey, author's interview, 16 July 1973.

8. Robert L. Rudd, *Pesticides and the Living Landscape.*

9. President's Science Advisory Committee, *Use of Pesticides.* U. S. Congress, Senate, Subcommittee on Reorganization and International Organization of the Committee on Government Operations, *Interagency Coordination on Environmental Hazards (Pesticides).*

10. The *Lucky Dragon* stirred a great many fears among Americans. The most striking indication is the upsurge in popular literature about fallout hazards. A sampling of the articles covered in the *Readers' Guide to Periodical Literature,* or a perusal of the titles in the

period 1953 to 1957, shows a marked and rapid increase in reports on this problem. On scientific opinion, the *Bulletin of the Atomic Scientists* is probably the best single source (particularly the November 1955 issue, which is largely about the effects of fallout on genes).

11. Again the popular literature, and letters to editors, are graphic indications of the concern felt by many Americans and people in other countries as well. A review of the scientific work on the subject in these years is J. Laurence Kulp, Walter R. Eckelmann, and Arthur R. Schulert, "Strontium-90 in Man." See also W. F. Libby, "Radioactive Fallout and Radioactive Strontium."

12. A speech Barry Commoner gave to the December 1957 meeting of the American Association for the Advancement of Science is an excellent description of his concerns: Barry Commoner, "The Fallout Problem" *Science*. See also Barry Commoner, *Science and Survival*.

13. *Washington Post*, 15 July 1962, 1. Insight Team of the Sunday Times of London, *Suffer the Children: The Story of Thalodimide*

14. William J. Darby, "A Scientist looks at *Silent Spring*," 2.

15. Pesticides and the Industry are Target of One-Sided Attack." "For the Defense: Pesticide Authorities Speak Out." "NACA Speakers Empahsize Industry's Role in Pesticide Safety." Robert H. White-Stevens, "Communications Create Understanding." On the whole environmental controversy, not just DDT, see Frank Graham Jr. *Since Silent Spring*.

16. 68 Stat 511. The amendment provided for a zero tolerance level for any material shown to be carcinogenic in man or animal. This requirement, which depended more on analytical chemistry than farm practice, proved difficult to enforce.

17. *New York Times*, 10 November 1959, 1:2.

18. Ibid.

19. Ibid.

20. *New York Times*, 11 November 1959, 1:4,5.

21. Ibid. 15 November 1959, 1:7,3.

22. The most accessible source of statements on these lines is that of Max Sobelman, comp. *Selected Statements from State of Washington DDT Hearings Held in Seattle, October 14, 15, 16, 1969*. On the practice of showing DDT was harmless by eating it: *Milwaukee Journal*, 10 June 1971; *Milwaukee Sentinel*, 2 October 1971. In author's interview, 1 March 1974, R. Keith Chapman admitted that he had also eaten DDT in front of classes to show that it was harmless.

23. See Ernest G. Moore, *The Agricultural Research Service*, 195. Whitney Gould, Madison newspaper reporter, said (author's interview, 6 July 1973) that she overheard economic entomologists muttering at the Wisconsin DDT hearing that the case was undermining their professional reputations. The same sentiments are also in the "Critique of 'Insecticides and People' " circulated by the Wisconsin Department of Entomology after the radio discussion of 1962.

24. Ira Baldwin, "Chemicals and Pests,"

25. Wisconsin State Broadcasting Service, "Summary of Broadcasts Dealing with *Silent Spring* by Rachel Carson and the Subject of Pesticides."

26. *Capital-Times*, 21 January 1963.

27. The author of the "Critique" remained anonymous for the entire controversy. R. Keith Chapman confirmed in an interview with the author, 1 March 1974 that he wrote the critique.

28. Entomology Department, "Critique," 1.

29. Ibid., 1, 14.

30. Ibid., 17.

31. Ibid.

32. American Association of Economic Entomologists, *AAEE—Presidential Addresses, 1889–1911*. The addresses provide an interesting self-portrait of the ideal economic entomologist and of the desired state of the profession.

33. Carson, *Silent Spring*, 261.

34. Darby, "Scientist Looks at *Silent Spring*," 2.

35. Donald Fleming has begun the task of setting out the ideological and intellectual underpinnings of the environmental movement in "Roots of the New Conservation Movement." Fleming's concentration on a few leaders, although useful, is not the last word on the subject. Donald Worster, *Nature's Economy*, presents another part of the puzzle and is well worth the time of any environmental historian. There is more to be said, though, and one hopes environmental historians will devote more time to the subject.

36. President's Science Advisory Council, *Use of Pesticides*. U. S. Senate, *Interagency Coordination*. National Research Council, *Pest Control and Wildlife Relationships*, parts 1–3.

37. President's Science Advisory Council, *Use of Pesticides*, 10–11.

38. Ibid., 16–19.

39. U. S. Senate, *Pesticides and Public Policy*, 66.

40. An early account of the effect of the new pesticides on control policy is G. C. Ulyett, "Insects, Man and the Environment."

41. National Research Council, *Pest-Control and Wildlife*, ii.

42. Roland C. Clement, "Review of *Pest Control and Wildlife Relationships*, part 3. Roland C. Clement, author's interview, 2 June 1977.

43. Weisner was, in fact, the first witness before the committee, U. S. Senate, *Interagency Coordination*.

44. Testimony of Rachel Carson in U. S. Senate, *Interagency Coordination*, 204–206.

45. Testimony of Orville Freeman in U. S. Senate, *Interagency Coordination*, 84–124.

46. U. S. Senate, *Interagency Coordination*, 269–270.

47. Testimony of George Lynn in U. S. Senate, *Interagency Coordination*, 334.

48. Quoted in U. S. Senate, *Interagency Coordination*, 65.

49. Ibid., 484–525, 689–703.   50. Ibid., 305.

51. Ibid., 410–412, 452.   52. Ibid., 131–133.

53. Ibid., 835–864.   54. Ibid., 570–576.   55. Ibid., 70.

56. Ibid., 1097–1165.   57. Ibid., 50.

58. Victor J. Yannacone, tape recording of press conference at house of Owen and Lorrie Otto, 8 October 1968. Recording from Lorrie Otto.

59. U. S. Senate, *Pesticides and Public Policy*, 66, 14.

60. Ibid., 6.

CHAPTER 6

1. Joseph J. Hickey and James E. Roelle, "Conference Summary and Conclusions." Charles F. Wurster, letter to Lorrie Otto, 27 February 1969. Letter is in the possession of Mrs. Otto.

2. Joseph J. Hickey and Daniel W. Anderson, "The Peregrine Falcon: Life History and Population Literature," 34.

3. Ibid., 8.

4. Joseph J. Hickey, author's interview, 16 July 1973. Derek A. Ratcliffe, "The Status of the Peregrine in Great Britian."

5. "Editorial," *Bird Study* 10 (June 1963), 55.

6. Daniel D. Berger, Charles R. Sindelar, Jr., and Kenneth E. Gamble, "The Status of Breeding Peregrines in the Eastern United States." Hickey interview.

7. Hickey and Anderson, "Peregrine Falcon: Life Study," 34.

8. Joseph J. Hickey, "Introduction" to Joseph J. Hickey, *Pergrine Falcon Population—Their Biology and Decline*.

9. Frank N. Hamerstrom, Jr., "An Ecological Appraisal of the Peregrine Decline," 509.

10. "Pesticides as Possible Factors Affecting Raptor Populations—Summary of a Round Table Discussion and Additional Comments by the Conferees," 419–448, and Eldridge G. Hunt, "Pesticide Residues in Fish and Wildlife of California," 455–460.

11. Tom J. Cade in a general discussion, "Population Biology and Significance of Trends," 548.

12. The proceedings of the Monks' Hole conference are in *Pesticides in the Environment and Their Effects on Wildlife, Journal of Applied Ecology* 3 (supplement). The Roland Clement papers, Audubon Society, contains notes and memoranda from Clement and Alexander Sprunt IV on the Port Clinton conference.

13. Clement papers, Audubon Society.

14. "Roundtable Discussion—Pesticides as Possible Factors Affecting Raptor Populations," 461–483. "General Statement by the Participants of the North Atlantic Treaty Organization Advanced Study Institute on Pesticides in the Environment," Appendix I in *Pesticides in the Environment, Journal of Applied Ecology* 3 (supplement), 297–298. Roland Clement, author's interview, 2 June 1977.

15. Hickey and Roelle, "Conference Summary," 563–564.

16. Hickey and Roelle, "Conference Summary," 565. Hickey interview.

17. Larry G. Hart, Robert W. Shultice, and James R. Fouts, "Stimulatory Effects of Chlordane on Hepatic Microsomal Drug Metabolism in the Rat."Larry G. Hart and James R. Fouts, "Effects of Acute and Chronic DDT Administration on Hepatic Microsomal Drug Metabolism in the Rat."

18. Testimony on this point was offered by Lucille F. Stickel in Wisconsin Department of Natural Resources, "In the Matter of the Petition of the CNRA of Wisconsin, Inc. et al. for a Declaratory Ruling on DDT," 1214–1319.

19. Testimony of Stickel in Wisconsin Department of Natural Resources, "In the Matter of DDT," 1214–1319.

20. Hickey and Roelle, "Conference Summary," 565–566.

21. Hickey interview.

22. Stanton J. Kleinert, Paul E. Degurse, Thomas L. Wirth, and Linda C. Hall, *DDT and Dieldrin Residues Found in Wisconsin Fishes from the Survey of 1966.*

23. See, for example, testimonies of Kenneth J. Macek, 968–1041, Robert L. Rudd, 1320–1451, and Robert W. Risebrough, 589–611, in

Wisconsin Department of Natural Resources, "In the Matter of DDT."

24. Carl W. Buchheister, "The Audubon Society and the Pesticides Controversy."

25. Carl W. Buchheister, "Meeting the Pesticides Problem."

26. Jamie L. Whitten, *That We May Live.*

27. The account of the EDF is based on the minutes of the meetings through the summer of 1969, kept by Carol Yannacone; on the memoranda, letters, and transcripts of conversations kept by Victor J. Yannacone; on interviews with Victor J. Yannacone, 10 December 1973 and 17 March 1977, Carol Yannacone, 31 May 1977, Charles F. Wurster, 21 December 1973; and on correspondence with the Yannacones, Wurster, Woodwell, Taromina, Puleston, and Cooley. Correspondence is in possession of author.

28. Yannacone interview, 21 December 1973. Myra Gelband, "Brookhaven Town Natural Resources Committee: A Call for Action." Ms. Gelband was one of the founders of the BTNRC.

29. Arthur Cooley to author, personal communication, 15 February 1977.

30. Carol Yannacone interview. Case was *Yannacone et al. v. Dennison et al.,* 285 N.Y. S. 2a 476 (New York).

31. Yannacone and Wurster interviews, 21 December 1973.

32. Yannacone interview, 21 December 1973.

33. *Yannacone v. Dennison.*

34. Yannacone interview, 21 December 1973. Charles F. Wurster, "Conserve America," undated mimeographed sheet from Clement papers.

35. Roland Clement, memorandum of October 6, 1967, with additons to 3 November 1967, Clement papers.

36. Incorporation papers and minutes of first meetings, copies from Yannacone papers.

37. Charles F. Wurster, letters to Lorrie Otto, 27 February 1968, 3 April 1968, letters in possession of Otto.

38. George M. Woodwell, memorandum of 10 January 1968, copy from Yannacone papers.

39. Charles F. Wurster to author, personal communication, December 1976.

40. Yannacone interview, 21 December 1973.

41. Yannacone interview, 21 December 1973.

42. Charles F. Wurster, speech in Milwaukee, October 1968. Tape recording from Lorrie Otto.

43. Yannacone interview, 21 December 1973.

44. Yannacone interview, 21 December 1973. *Environmental Defense Fund v. Michigan Department of Agriculture, the incorporated Cities of Fremont, Muskegon, Breenville, Rockford, East Lansing, Lansing, East Grand Rapids, Holland, and the Incorporated Village of Spring Lake, all in the State of Michigan; Detroit Free Press*, 4 November 1967; *Kalamazoo Gazette*, 4 November 1967. Newspaper clippings and transcripts from papers of Victor J. Yannacone, Jr.

45. *Environmental Defense Fund, Inc., Plaintiff, v. B. Dale Ball, Director; Dean Lovitt, Chief, Plant Industry Division; Donald White, Regional Supervisor, Plant Industry Division, Michigan Department of Agriculture.*

46. Victor J. Yannacone, Jr., talk with reporters, 8 October 1968 at Otto's house, tape from Lorrie Otto.

47. *Environmental Defense Fund v. B. Dale Ball* (1968), 1–20.

48. Ibid., 22–24.   49. Ibid., 29.   50. Ibid., 39–68.

51. Ibid., 83–96.   52. Ibid., 87–121.

53. Otto interview, 19 November 1974.

54. Material on early activities of Citizens' Natural Resources Association of Wisconsin, Inc. from flyers and bulletins in the Scott papers.

55. Otto interview, 19 November 1974.

56. Otto interview, 19 November 1974.

57. Wiconsin Department of Natural Resources, "Fredrick M. Baumgartner et al. v. City of Milwaukee and Buckley Tree Service," docket 4C-68-5-1 (unpublished transcript, Wisconsin Department of Natural Resources, 1968), hearing held 18 October 1968.

58. Wisconsin Department of Natural Resources, "Baumgartner v. Milwaukee," 11.

59. Maurice Van Susteren, author's interview, 6 June 1973.

60. Van Susteren provided the example to the EDF; confirmed by both Yannacone and Wurster in author's interview.

CHAPTER 7

1. Aldo Leopold, *A Sand County Almanac*, 188.

2. William Reeder, author's interview, 13 September 1973; Lorrie Otto, author's interviews, 18 February 1974 and 19 November 1974; Victor Yannacone, author's interview, 21 December 1973; Charles F. Wurster, memorandum to prospective witnesses, 22 November 1968, copy from Yannacone papers.

3. Reeder interview.

4. Wurster memorandum, 22 November 1968.

5. Louis A. McLean, author's interview, 5 November 1973.

6. In undated interviews, late 1974, Van Susteren pointed to McLean's qualifications and his expertise at lobbying. Yannacone, author's interview, 17 March 1977, said that he had hoped to catch the industry lawyers off balance and by surprise.

7. McLean interview.

8. Citations in parentheses are to pages of the official transcript of the hearing. Wisconsin Department of Natural Resources, "In the Matter of the Petition of the CNRA of Wisconsin, Inc. et al. for a Declaratory Ruling on DDT."

9. (Madison) *Capital-Times*, 2 December 1968.

10. Transcript of hearing, 367; *Capital-Times*, 4 December 1968; (Madison) *Wisconsin State Journal*, 5 December 1968.

11. Louis A. McLean, "Pesticides and the Environment." Introduced into the Wisconsin DDT hearing as exhibit 1, 2 December 1968.

12. Harmon Henkin, Martin Merta, and James Staples, *The Environment, the Establishment, and the Law*, 14.

13. Hugh H. Iltis, author's interview, 19 June 1973.

14. Henkin et al. *The Environment, the Establishment, and the Law*.

15. Ibid., 14.

16. Sorption is a general term for the processes of absorption and adsorption.

17. See, for example, the comments of Samuel Rotrosen, general manager of Montrose Chemical Corporation of California in the *Wall Street Journal*, 4 March 1969.

18. *Milwaukee Journal*. 4 December 1968.

19. G. C. Ullyet, "Insects, Man, and the Environment," 462. See also Chapter 1.

20. Paul DeBach, "The Necessity of an Ecological Approach to Pest Control on Citrus in California."

21. See Appendix G—Vapor Phase Chromatography.

22. *Milwaukee Journal*, 7 December 1968, *Capital-Times*, 6 December 1968.

23. Joseph J. Hickey, J. A. Keith, and Francis B. Coon, "An Exploration of Pesticides in a Lake Michigan Ecosystem."

24. Yannacone interview, 17 March 1977.

25. Wisconsin Department of Agriculture and the University of Wisconsin, "Dutch Elm Disease Control Recommendations for 1969."

26. Letter from the Chilton Rotary Club to L. P. Voight, secretary of the Wisconsin Department of Natural Resources, 11 December 1968, Scott papers.

27. "People vs. DDT," a flyer distributed by the Conservation Research and Action Project of the Science Students Union, University of Wisconsin, about 11 December 1968, Scott papers.

28. *Wisconsin State Journal,* 13 December 1968; *Milwaukee Sentinel,* 13 December 1968.

29. Van Susteren interview, 6 June 1973.

30. *Capital-Times,* 14 January 1969; *Wisconsin State Journal,* 15 January 1969; *New York Times,* 15 January 1969.

31. Van Susteren interview, 6 June 1973.

CHAPTER 8

1. *Janesville [Wisconsin] Daily Gazette,* 29 January 1969, clipping from Scott papers.

2. Wisconsin Legislative Council Staff "Regulation of DDT and Other Pesticides: Recent Developments in Wisconsin." 5, copy from Scott papers.

3. Wisconsin Legislative Reference Bureau, drafting file 1969—assembly bill 819, copy from the Legislative Reference Bureau.

4. *New York Times,* 30 April 1969, 43:1.

5. Hal Higdon, "Obituary for DDT (in Michigan)." The fish, condemned for human consumption, eventually went into mink food. It apparently produced poisoned milk and other reproductive problems and the fur farmers sued the suppliers. A Calumet County, Wisconsin, jury awarded the farmers damages (*Valiga v. National Food Co.* 58 Wis2d 232, decided 20 April 1973), a judgment upheld by the Wisconsin Supreme Court (*Milwaukee Journal,* 23 May 1973 and decision cited above.)

6. Wisconsin Resource Conservation Council, untitled handbill; *Eagle River—Vilas County News Review,* 10 April 1969; *Milwaukee Journal,* 11 April 1969; (Madison) *CapitalTimes,* 11 April 1969, 19 April 1969; all matrials are from Scott papers.

7. Correspondence of the EDF in papers of Victor J. Yannacone, Jr. Charles F. Wurster and Victor J. Yannacone, author's interviews, 21 December 1973.

8. Charles F. Wurster, memorandum of 11 August 1969 to all EDF Trustees, p. 3, copy from Yannacone papers.

9. U. S. Congress, Senate, Subcommittee on Reorganization and International Organization of the Committee on Government Operations, *Interagency Coordination in Environmental Hazards (Pesticides)*, 88th Congress, 1st Session.

10. *Capital-Times*, 29 April 1969.

11. Environmental Protection Agency, "Public Hearing on DDT," 4587–4589.

12. Louis A. McLean, author's interview, 5 November 1973.

13. Yannacone interview. *New York Times*, 28 February 1968, 35:2.

14. *Milwaukee Journal*, 1 May 1969.

15. Testimony of Jesse Steinfeld in Environmental Protection Agency, "Public Hearing On DDT," 1367.

16. Brief submitted by petitioners in support of the original petition to have DDT declared a pollutant, 15 September 1969; action was Department of Natural Resources, "In the Matter of the Citizens' Natural Resources Association of Wisconsin, Incorporated et al. for a Declaratory Ruling on DDT" copy from the Scott papers.

17. Hugh H. Iltis, author's interview, 19 June 1973, confirmed by Charles F. Wurster, December, 1976, personal communication.

18. *Capital-Times*, 14 January 1969.

19. *New York Times*, 29 March 1970, IV 6:3.

20. Francis B. Coon, author's interview, 23 October 1973; Ellsworth Fisher, author's interview, 15 April 1974.

21. PCB is an abbreviation for polychlorinated biphenyls, a class of compounds with industrial uses. Although discussion of their appearance in gas chromatography focused on one peak, interfering with or lying close to DDT and DDD, PCBs display a group of peaks, differing for each formulation. See Appendix G—Vapor Phase Chromatography.

22. Testimony of George J. Wallace in Wisconsin Department of Natural Resources, "In the Matter of the Petiton of the CNRA of Wisconsin, et al. for a Declaratory Ruling on DDT," 2521–2524.

23. Roland Clement C., "The Pesticides Controversy" and "The Report on Pesticide Use." Author's interview, 2 June 1977.

24. U. S. Congress, Senate, Subcommittee on Reorganization and International Organization of the Committee on Government Operations, *Interagency Coordination*, 1st Session.

25. Theodore Goodfriend, author's interview, 9 July 1973.

26. Maurice Van Susteren, author's interview, 6 July 1973, and subsequent conversations during 1974 and early 1975, undated.

CHAPTER 9

1. EDF correspondence, papers of Victor J. Yannacone, Jr.
2. Paul Erlich, *The Population Bomb.*
3. The new book by the EDF and Robert H. Boyle, *Malignant Neglect* (New York: Knopf, 1979), is the best summary of connections between the environmentalists and the public health movement. Joseph L. Sax, *Defending the Environment,* provides some interesting material on the early work in environmental law.
4. The literature on the Noble Savage and on aboriginal man-land relationships is large and growing, particularly on issues such as hunting and the extinction of North American game in the late Pleistocene period. An interesting recent study (particularly the epilogue) is Calvin Martin, *Keepers of the Game: Indian-Animal Relationships and the Fur Trade.*
5. *Washington Post,* 22 April 1970, 23 April 1970. Ronald Lee Shelton, "The Environmental Era," 327–332, provides a short account. Like many other aspects of recent environmental history, Earth Day awaits serious study.
6. Government Accounting Office, "Need to Improve Regulatory Enforcement Procedures Involving Pesticides," 141–180, and "Need to Resolve Questions of Safety Involving Certain Registered Uses of Lindane Pesticide Pellets," 181–217. U. S. Congress, House, Committee on Government Operations, *Deficiencies in Administration of Federal Insecticide, Fungicide, and Rodenticide Act.* The Mrak Commission report is more formally: U. S. Department of Health, Education, and Welfare, *Report of the Secretary's Commission on Pesticides and Their Relationship to Environmental Health,*
7. Government Accounting Office, "Need to Improve" and "Need to Resolve."
8. U. S. Congress, House, Committee on Government Operations, *Deficiencies in Administration,* both the hearings and the report.
9. For a discussion of this point, see Mrak report, 471.
10. EDF press release, 7 October 1969; *Washington Post,* (editorial), 25 October 1969; *New York Times,* 4 November 1969. All material from Yannacone papers.
11. Mrak report, 471, 482.   12. Ibid., 493–498.
13. Environmental Defense Fund, "Memorandum of Understanding" (between Yannacone and the other EDF trustees), 18 April 1969; Minutes of EDF meeting, 24 April 1969; Wurster, memorandum to EDF trustees, 11 August 1969.

14. Bylaws of the Environmental Defense Fund, June 1969; Victor Yannacone, author's interview, 21 December 1973; Carol Yannacone, author's interview, 31 May 1977; Charles F. Wurster, author's interview, 21 December 1973.

15. Yannacone and Wurster interviews; Charles F. Wurster, personal communication, December 1976, in possession of author.

16. Charles F. Wurster, memorandum to EDF Trustees, 4 June 1969, 2 July 1969, 11 August 1969.

17. Arizona Board of Pesticides Control, rule no. 20, effective 13 January 1969, from Scott papers. *National Observer*, 16 June 1969; *State Journal* (Lansing, Michigan), 17 April 1969, 18 June 1969, 19 June 1969; *Chicago Sun-Times*, 13 June 1969; *New York Times*, 20 April 1969; (Madison) *Capital-Times*, 18 July 1969, 19 July 1969; *Wisconsin State Journal*, 20 August 1969; all clippings from Scott papers. "The Non-Ban on DDT—A Federal Phase-Out Fizzle." Audubon press release, 25 June 1969, Clement papers.

18. EDF letter to Mrs. Archie Carr, Mycanopy, Florida, 23 June 1969; EDF letter to Elvis Stahr, National Audubon Society, 9 June 1969; *EDF v. Corps of Engineers et al.*, suit filed in U. S. District Court for the District of Columbia, 16 September 1969, minutes, 15 August 1969; *EDF v. Hoernor-Waldorf Corporation*, Civil Action 1644, U. S. District Court, District of Montana, Missoula Division; all materials from Yannacone papers.

19. "Petition Requesting the Suspension and Cancellation of Registration of Economic Poisons Containing DDT," Environmental Defense Fund, Sierra Club, West Michigan Environmental Action Council and National Audubon Society, petitioners, to Clifford M. Hardin, Secretary of Agriculture, copy from Clement papers; EDF newsletter, 8 January 1970; EDF press release, 7 October 1969.

20. *New York Times*, 28 February 1968, 35:2.

21. EDF press release, 7 October 1969.

22. *New York Times*, 13 November 1969, 1:8; 21 November 1969, 1:4.

23. *New York Times*, 30 December 1969, 14:1.

24. *New York Times*, 29 March 1970, IV, 6:3.

25. 428 *Federal Reporter* 2nd series, 1096–1099, copy from Clement papers.

26. EDF newsletter, 8 January 1970, 1.

27. Section 102 has become a lawyer's dream and an adminis-

trator's nightmare. There is no doubt that the requirement of an environmental impact statement has made some agencies more conscious of their responsibilities to the environment and made all of them more vulnerable to citizen action. A full assessment of the impact of the bill will be a long time in coming, for the act continues to generate cases, precedents, and decisions.

28. For an introduction to the problems and possibilities of the act, see Richard N. L. Andrews, *Environmental Policy and Administrative Change* and Frederick H. Anderson, *NEPA in the Courts*.

29. Whitten has a reputation among environmentalists as one of their inveterate foes, and Audubon commented on the new powers of the subcommittee in "Non-Ban on DDT." The best insight into Whitten's own ideas is his defense of DDT, *That We may Live*.

30. *New York Times*, 8 January 1971, 1:5. A study that appeared too late to be used here but that sheds additional light on the legal aspects of the case in Angus A. MacIntyre's "The Politics of Nonincremental Domestic Change: Major Reform in Federal Pesticide and Predator Control Policy," Ph.D. dissertation, University of California, Davis, 1980.

31. "Non-Ban on DDT."

32. Numbers in parentheses in this chapter refer to pages in the transcript of the hearing held by the Environmental Protection Agency, "Public Hearing on DDT."

33. Testimony of Jesse Steinfeld, 1367; Marshall Laird, 60–91.

34. Testimony of Norman Borlaug, 2597–2665. Borlaug was clearly committed not just to DDT but to the entire complex of mechanized, high-technology agriculture.

35. Saffiotti's attack, one of the strongest made against the medical studies of DDT's defenders was, significantly, made by a witness, not, as has been the case in Madison, by a lawyer on cross-examination.

36. Testimony of Umberto Saffiotti, 4002–4077.

37. The transcript contains a paging error, jumping from 3099 to 4000, and only returning to a correct numbering several hundred pages later. Numbers cited in the text are those in the transcript; anyone using the transcript of the hearing is advised to be wary.

38. Testimony of William L. Reichel, 4032–4048; Jerry A. Burke, 4049–4074; Daniel L. Stalling, 4075–5022. Actual number at 5022 is 3222, due to paging errors. Testimony of H. Page Nickolson, 3227–3278; Tony J. Peterle, 3282–3355; John B. Dimond, 3460–3519; Phillop C. Kearney, 3519–3566; Charles D. Gish, 3567–3691.

39. Testimony of Joseph C. Headley, 6167–6228.
40. Robert Gillette, "DDT: In Field and Courtroom a Persistent Pesticide Lives On."
41. Transcript of hearing, 6492–6514.
42. Testimony of Lawrence R. Cory, 7050–7118; George M. Woodwell, 7193–7228; Alden Dexter Hinckley, 7251–7282; Samuel S. Epstein, 7310–7404, Jerry Mosser, 7500–7540; Frederick W. Plapp, 7639–7707.
43. William A. Butler, "Exceptions of Intervenors Environmental Defense Fund, Inc., National Audubon Society, Western Michigan Environmental Action Council, and Sierra Club to the Hearing Examiner's Recommended Decision and Conduct of the Hearing," in RE Consolidated DDT Hearings, filed 12 May 1972, III–268.

EPILOGUE

1. Edmund M. Sweeney, "Hearing Examiner's Recommended Findings, Conclusions, and Orders (40 DFR 164)," copy from EDF files.
2. Van Susteren, author's interview, 6 June 1973, said that the opinion was delayed due to the insistence of his superiors in the Department of Natural Resources. It is difficult either to find out what happened during this period or to print it, for Van Susteren received what he said was a demotion and what his superiors characterized as a lateral promotion shortly after hearing the case. He filed at least five separate legal actions against the department. The internal history of the DNR will probably be an excellent subject for a dissertation, but not for some years to come.
3. Maurice Van Susteren, "Examiner's Summary of Evidence and Proposed Ruling," 27, 28, copy from Scott papers.
4. Willard S. Stafford, "Exceptions to Examiner's Statement of Statutes, Rules, and Issues"; Maurice Van Susteren, "Examiner's Ruling on Motion of Industry Task Force for DDT of the National Agriculture Chemicals Association in Support of Motion to Dismiss Proceedings"; Wisconsin Department of Natural Resources. "Order Affirming and Adopting Examiner's Proposed Rulings," 23 October 1970; all copies from Scott papers. Norris Maloney, ruling rejecting the Industry Task Force for DDT's request for circuit court review of DDT ruling and order, "The Industry Task Force for DDT of the NACA v. DNR," 4 November 1971, copy from EDF files.

5. James G. Hilton (head, DDT Advisory Committee), "Report of the DDT Advisory Committee to William D. Ruckelshaus, Administrator, Environmental Protection Agency," 9 September 1971, copy from EDF files.

6. U. S. Department of Health, Education and Welfare, *Report of the Secretary's Commission on Pesticides and their Relationship to Environmental Health*.

7. Environmental Protection Agency, "Consolidated DDT Hearings, Opinion and Order of the Administrator."

8. The last exemption proved to be a controversial one; environmentalists claimed that the tussock moth's population would naturally fall due to epidemic disease; loggers and the pesticide industry claimed that there was a real need for the chemical. Industry used the case, so the environmentalists thought, as an opening wedge in a campaign to bring DDT back. For a sampling of opinion, see U. S. Congress, House Subcommittee on Forests of the Committee on Agriculture, *Permit the Use of DDT*.

9. Accounts of the USDA's effectiveness are in Government Accounting Office, "Need to Improve Regulatory Enforcement Procedures Involving Pesticides" and "Need to Resolve Questions of Safety Involving Certain Registered Uses of Lindane Pesticide Pellets."

10. On the Congressional study of a new FIFRA, see U. S. Congress, House, Committee on Agriculture, *Federal Pesticide Control Act of 1971*, and U. S. Congress, Senate, Subcommittee on Agricultural Research and General Legislation of the Committee on Agriculture and Forestry, *Federal Environmental Pesticide Control Act*

11. EDF newsletter, 1970–1977; *Washington Post*, 5 May 1979; Environmental Defense Fund and Robert H. Boyle, *Malignant Neglect*, 130; National Wildlife Federation, *Conservation Newsletter*, January 1975.

12. Environmental Defense Fund, "Newsletter," January 1975.

13. Environmental Defense Fund and Boyle, *Malignant Neglect*.

14. Robert van den Bosch, *The Pesticide Conspiracy*.

15. The development of economic entomology since World War II is the subject of a forthcoming manuscript by John Perkins, Miami University, Oxford, Ohio, tentative title: *Insects and Experts: The Political Economy of Scientific Change*. Perkins's work will, I hope, be a definitive treatment; I feel it useless to do more than mention the subject.

16. A short introduction to lines of research of the early 1970s is in

U. S. Congress, Senate, Committee on Agriculture and Forestry, *Pest Control Research*. See also John H. Perkins, "The Quest for Innovation in Agricultural Entomology, 1945–1978."

17. See van den Bosch, *Conspiracy*.

APPENDIX B

1. Donald E. H. Frear, *Chemistry of Insecticides, Fungicides, and Herbicides* (New York: D. Van Nostrand, 1948), 15.

APPENDIX C

1. Edwin P. Laug, Frieda M. Kunze, and C. S. Prickett, "Occurrence of DDT In Human Fat and Milk."

2. Testimony of Wayland J. Hayes, Jr. in U. S. Congress, House, House Select Committee to Investigate the Use of Chemicals in Food Products, *Chemicals in Food Products*, 82nd Congress, 1st Session, 101.

3. "Summary of Investigations of DDT Residues on Food and DDT Storage in Human Fat—Technical Development Branch, Communicable Disease Center, Public Health Service, Savannah, Georgia," in *Chemicals in Food Products*, 82nd Congress, 1383–1384.

4. Wayland J. Hayes, Jr., "The Effect of Known Repeated Oral Doses of Chlorophenothane (DDT) in Man"; Wayland J. Hayes, Jr., Griffith E. Quimby, Kenneth C. Walker, Joseph W. Elliot, and William M. Upholt, "Storage of DDT and DDE in People with Different Degrees of Exposure to DDT," *A.M.A. Archives of Industrial Health*, 17 (November 1958), 398–400.

5. Storage is neither uniform among the population nor unaffected by environmental influences. Vegetarians store less DDT than the general population, since animal fat is a primary source of DDT residues; breast-fed babies store more.

APPENDIX D

1. Harvey W. Wiley, *The History of a Crime Against the Food Law*.

2. Oscar E. Anderson, *The Health of a Nation: Harvey W. Wiley and the Fight for Pure Food*.

APPENDIX E

1. The list of manufacturers from 1969 is from the testimony of
Samuel Rotrosen in the Wisconsin Department of Natural Re-
sources, "In the Matter of the Petition of the Citizens' Natural
Resources Association of Wisconsin, Inc. et al. for a Declaratory
Ruling on DDT," 2468.

APPENDIX G

1. Milton S. Schecter, S. B. Solway, Robert A. Hayes, and H. L.
Haller, "Colorimetric Determination of DDT: Color test for Related
Compounds," *Industrial and Engineering Chemistry, Analytical Edition*
18 (November 1945), 705.

2. Ibid., 708.

BIBLIOGRAPHY

THIS LIST includes all the printed or easily available sources cited in the book. It does not include most ephemera such as handbills, occasional memoranda, or letters, cited a single time, or personal correspondence directed to the author. In addition to the works cited below, seven tape-recorded interviews with participants in the Wisconsin DDT hearing are available at the State Historical Society of Wisconsin, 816 State Street, Madison, Wisconsin. They are with Joseph J. Hickey, Hugh Iltis, Orie Loucks, Lorrie Otto, Maurice Van Susteren, Charles Wurster, and Victor Yannacone.

American Association of Economic Entomologists. *AAEE—Presidential Addresses, 1889–1911*. No publisher, no date.
American Medical Association. "Dangers of Lead Arsenate as Spray in Orchards." *Journal of the American Medical Association* 105 (17 August 1935), 531.
———. "Hazards of Contaminated Fruits and Vegetables." *Journal of the American Medical Association* 109 (2 July 1937), 135.
———. "Health Hazards of Pesticides." *Journal of the American Medical Association* 137 (28 August 1948), 1603.
———. "Pesticides: Chemical Contaminants of Foods." *Journal of the American Medical Association* 138 (28 August 1948), 1604–1605.
———. "Spray Residues on Foods." *Journal of the American Medical Association* 108 (April 1937) 1178.
American Medical Association, Council on Foods and Nutrition. "Health Hazards of Pesticides." *Journal of the American Medical Association* 137 (28 August 1948), 1604.
Anderson, Frederick H. *NEPA in the Courts*. Washington: Resources for the Future, 1973.
Anderson, Oscar E. *The Health of a Nation: Harvey W. Wiley and the Fight for Pure Food*. Chicago: University of Chicago Press, 1958.
Andrews, Richard N. L. *Environmental Policy and Administrative Change*. Lexington, Mass.: D. C. Heath, 1976.
Baker, John H. "The President's Report to You." *Audubon Magazine* 47 (September–October 1945), 309–315.
Baker, John H. "The President Reports to You." *Audubon Magazine* 60 (March–April 1958), 78.

Baldwin, Ira. "Chemicals and Pests." *Science* 137 (28 September 1962), 1041–1043.

Barker, Roy J. "Notes on Some Ecological Effects of DDT Sprayed on Elms." *Journal of Wildlife Management* 22 (July 1958), 269–274.

Berger, Daniel D., Charles R. Sindelar, Jr., and Kenneth E. Gamble. "The Status of Breedings Peregrines in the Eastern United States." In Hickey. *Peregrine Falcon Populations*, 165–173.

Bishopp, Fred C. "Insect Problems in World War II with Special Reference to the Insecticide DDT." *American Journal of Public Health* 35 (April 1945), 373–378.

Brittin, W. A. "Chemical Agents in Food." *Food Drug Cosmetic Law Journal* 5 (September 1950), 590–597.

Britton, W. E. "The Academic Training of the Entomologists in Colleges and Experiment Stations of the United States." *Journal of Economic Entomology* 8 (February 1915), 72–78.

Broley, Charles L. "The Plight of the American Bald Eagle." *Audubon Magazine* 60 (July–August 1958), 162.

Buchheister, Carl W. "The Audubon Society and the Pesticides Controversy." Speech to the Society of American Foresters, 17 March 1965. Copy from Clement papers.

————. "Four of Six Conservation Goals Reached." *Audubon Magazine* 64 (January–February 1962), 34.

————. "Meeting the Pesticides Problem." Speech, 2 October 1962. Copy from Clement papers.

Burgess, A. F. "Gipsy Moth Barrier Zone Maintenance Problems." *Journal of Economic Entomology* 23 (August 1930), 720–725.

————. "The Present Status of the Control of the Gypsy Moth and the Brown-Tail Moth by Means of Parasites." *Journal of Economic Entomology* 19 (April 1926), 289–294.

————. "The Value of Uninfested States of Gypsy Moth Control and Extermination." *Journal of Economic Entomology* 33 (June 1940), 558–561.

Carson, Rachel. "A Report on Progress." In Paul Brooks, *The House of Life*. New York: Houghton Mifflin, 1972. Pp. 243–246.

————. *Silent Spring*. Boston: Houghton Mifflin, 1962.

Carter, Luther J. "Conservation Law (I): Seeking a Breakthrough in the Courts." *Science* 166 (19 December 1969), 1487–1491.

————. "Conservation Law (II): Scientists Play a Key Role in Court Suits." *Science* 166 (26 December 1969), 1601–1606.

——. "DDT: The Critics Attempt to Ban its Use in Wisconsin." *Science* 163 (7 February 1969), 548–551.

——. "Environmental Law (I): Maturing Field for Lawyers and Scientists." *Science* 179 (23 March 1973), 1205–1209.

——. "Environmental Law (II): A Strategic Weapon Against Degradation?" *Science* 179 (30 March 1973), 1310–1311.

——. "Environmental Pollution: Scientists Go to Court." *Science* 163 (22 December 1967), 1552.

——. "Pesticides: Environmentalists Seek New Victory in Frustrating War." *Science* 181 (13 July 1973), 143–144.

Case, R. A. M. "Toxic Effects of 2, 2bis (p-chlorpheny) 1,1,1 trichloroethane (DDT) in Man." *British Medical Journal* 15 December 1945, 842–845.

Cavers, David F. "The Food Drug and Costmetic Act of 1938: Its Legislative History and its Substantive Provisions." *Law and Contemporary Problems* 6 (Winter 1939), 2–42.

Centennial Committee: Gladys Baker, Wayne D. Rasmussen, Vivian Wiser, and Jane M. Porter. *A Century of Service: The First 100 Years of the USDA*. Washington: Government Printing Office, 1963.

"Chemicals in Food." *Agricultural Chemicals* 7 (February 1952), 57, 59.

Clement, Roland C. "The Pesticides Controversy." *Environmental Affairs* 2 (Winter, 1972), 445–468.

——. "The Report on Pesticide Use." *Natural Resources Journal* 4 (October 1964), 247–251.

——. "Review of *Pest Control and Wildlife Relationships* Part III." *Audubon Magazine* 66 (January–February 1964), 60.

Clement papers—papers of Roland Clement, Vice-President of the National Audubon Society. Arthur A. Allen Archives, Laboratory of Ornithology, Cornell University, Ithaca, N.Y.

Coad, B. R. "Dusting Cotton from Airplanes." USDA Department Bulletin 1204. Washington: Government Printing Office, 1924.

——. "Killing Boll Weevils with Poison Dust." In United States Department of Agriculture. *U. S. Department of Agriculture Yearbook for 1920*. Washington: Government Printing Office, 1920. 241–253.

——. "Recent Experimental Work on Poisoning Cotton-Boll Weevils." USDA Department Bulletin 731. Washington: Government Printing Office, 1918.

Commoner, Barry. "The Fallout Problem." *Science* 127 (2 May 1958), 1023–1026.
———. *Science and Survival*. New York: Viking, 1967.
Comstock, John Henry. "The Present Methods of Teaching Entomology." *Journal of Economic Entomology* 4 (February 1911), 53–63.
Cottam, Clarence, and Elmer Higgins. "DDT and its Effect on Fish and Wildlife." *Journal of Economic Entomology* 39 (February 1946), 44–52.
Cox, H. E. "The Occurrence of Arsenic in Apples." *Analyst* 51 (1926), 132–137.
Cristol, Stanley J. and H. L. Haller. "The Chemistry of DDT—A Review." *Chemical and Engineering News* 23 (25 November 1945), 2070–2075.
"DDT Considered Safe for Insecticidal Use." *American Journal of Public Health* 34 (December 1944), 1312–1313.
Darby, William J. "A Scientist looks at *"Silent Spring."* Copyright American Chemical Society, 1962, copy from Aaron Ihde.
DeBach, Paul. *Biological Control by Natural Enemies*. Cambridge: Cambridge University Press, 1974.
———. "The Necessity for an Ecological Approach to Pest Control on Citrus in California." *Journal of Economic Entomology* 44 (August 1951), 443–447.
Decker, George C. "Agricultural Applications of DDT, with Special Reference to the Importance of Residues." *Journal of Economic Entomology* 39 (October 1946), 557–562.
Department of Lands and Forests, Ontario, Canada. *Forest Spraying and Some Effects of DDT*. Biological bulletin no. 2 Division of Research, 1949.
DeWitt, James B. "Chronic Toxicity to Quail and Pheasants of Some Chlorinated Insecticides." *Journal of Agricultural and Food Chemistry* 4 (October 1956), 863–866.
———. "Effects of Chemical Sprays on Wildlife." *Audubon Magazine* 60 (March–April 1958), 70–71.
———. "Effects of Chlorinated Hydrocarbon Insecticides upon Quail and Pheasants." *Journal of Agricultural and Food Chemistry* 3 (August 1955), 672–676.
Downs, Eldon W., and George F. Lemmer. "Origins of Aerial Crop Dusting." *Agricultural History* 39 (1965), 123–135.
Draize, John H., Geoffery Woodard, O. Garth Fitzhugh, Arthur A. Nelson, R. Blackwell Smith, Jr., and Herbert O. Calvery.

"Summary of Toxicological Studies of the Insecticide DDT." *Chemical and Engineering News* 22 (10 September 1944), 1503–1504.

Dunbar, Paul A. "The Food and Drug Administration Looks at Insecticides." *Food Drug Cosmetic Law Journal* 4 (June 1949), 233–239.

——. "Memories of Early Days of Federal Food and Drug Law Enforcement." *Food Drug Cosmetic Law Journal* 14 (February 1959), 87–138.

Dunlap, Thomas R. "The Triumph of Chemical Pesticides in Insect Control, 1890–1920." *Environmental Review*, no. 5 (1978), 38–47.

Dupree, Anderson Hunter. *Science in the Federal Government*. Cambridge: Belknap Press of Harvard University Press, 1957.

"Dutch Elm Disease Found Here." *Official Record: United States Department of Agriculture* 9 (4 September 1930), 3.

"Editorial." *Bird Study* 10 (June 1963), 55.

Environmental Defense Fund and Robert H. Boyle. *Malignant Neglect*. New York: Knopf, 1979.

Environmental Defense Fund, Inc., Plaintiff v. B. Dale Ball, Director, Dean Lovitt, Chief, Plant Industry Division, Donald White, Regional Supervisor, Plant Industry Division, Michigan Department of Agriculture. Civil Action no. 68-C-289, U. S. District Court, Eastern District of Wisconsin, 9 October 1968. Transcript of action from Scott papers.

Environmental Defense v. Michigan Department of Agriculture, the Incorporated Cities of Fremont, Muskegon, Breenville, Rockford, East Lansing, Lansing, East Grand Rapids, Holland, and the Incorporated Village of Spring Lake, all in the State of Michigan. Action no. 5760, U. S. District Court, Western District of Michigan, Southern Division, 3 November 1967.

Environmental Protection Agency. *DDT: A Review of Scientific and Economic Aspects of the Decision to Ban its Use as a Pesticide*. Washington: Government Printing Office, 1975.

——. "Public Hearing on DDT." Transcript of tape recording of hearing furnished to respondent EDF (hearing also called Consolidated DDT Hearing).

Ackerley, Robert L., and Charles A. O'Conner, III. "Group Petitioners' Analysis of the Evidence in the DDT Cancellation Proceedings." 2 March 1972.

Environmental Defense Fund: Butler, William A. "Exceptions of

Intervenors Environmental Defense Fund, Inc., National Audubon Society, Western Michigan Environmental Action
Council, and Sierra Club to the Hearing Examiner's Recommended Decision and Conduct of Hearing." In RE Consolidated DDT Hearings, filed 12 May 1972.

Environmental Defense Fund: Butler, William A. and John F.
Dienelt. "Summary Statement of Law and Analysis of Evidence." Summary and analysis in compliance with examiner's
ruling of 17 February 1972, submitted 2 March 1972.

Environmental Protection Agency. "Consolidated DDT Hearings, Opinion and Order of the Administrator." *Federal Register*
37 13369–13376.

Environmental Protection Agency: Fielding, Blaine and John C.
Kolojeski. "Respondent's Analysis of the Evidence, Summary
of the Law and Argument." No date.

Edmund M. Sweeney. "Hearing Examiner's Recommended
Findings, Conclusions and Orders (40 CFR 164, 32), Consolidated DDT Hearing."

Erlich, Paul. *The Population Bomb* New York: Ballantine—Sierra
Club, 1968.

Fernald, C. H. "The Evolution of Economic Entomology." In
American Association of Economic Entomologists. *AAEE—
Presidential Addresses 1889–1911*.

Fisher, Carl D. "Pesticides, Past, Present and Future, I. Insecticides, Miticides, and Rodenticides." *Chemical Week* 79 (27
October 1956), 59–90.

Fleming, Donald. "Roots of the New Conservation Movement."
Perspectives in American History 6 (1972), 7–91.

Folger, J. C. "The Commerical Apple Industry in the United
States." In U. S. Department of Agriculture. *Yearbook of Agriculture, 1918*. Washington: Government Printing Office, 1918.

"For the Defense: Pesticide Authorities Speak Out." *Agricultural
Chemicals* 17 (September 1962), 26.

Forbes, Stephen A. "The Ecological Foundations of Applied Entomology." *Annals of the Entomological Society of America* 8
(1915), 1–19.

Forbush, Edward H. and Charles H. Fernald. *The Gypsy Moth*.
Boston: Wright and Potter, state printers, 1896.

Frisbie, Walter S. "Federal Control of Spray Residues on Fruits and
Vegetables." *American Journal of Public Health* 26 (April 1936),
369–373.

Galloway, Beverley T. "The Grape Scare in New York." In U. S. Department of Agriculture. *Annual Report of the Secretary of Agriculture, 1891*. Washington: Government Printing Office, 1892. Pp. 375–376.

Geagley, W. C. "State Control of Spray Residues." *American Journal of Public Health* 26 (April 1936), 374–376.

Geer, Harriet A., Randall Lotta, and Arthur W. Lindquist. "The Dispersal of Insecticides." In U. S. Office of Scientific Research and Development. *Advances in Military Medicine*. Vol. 2. Pp. 636–645.

Gelband, Myra. "Brookhaven Town Natural Resources Committee: A Call for Action." Unpublished paper, 16 May 1969, copy from Art Cooley.

"General Statement by the Participants of the North Atlantic Treaty Organization Advanced Study Institute on Pesticides in the Environment." *Journal of Applied Ecology* 3 (supplement), 297–298.

George, John L. *The Program to Eradicate the Imported Fire Ant*. New York: The Conservation Foundation, 1958.

Gillette, Robert. "DDT: In Field and Courtroom a Persistent Pesticide Lives On," *Science* 174 (10 December 1971), 1108–1110.

Government Accounting Office. "Need to Improve Regulatory Enforcement Procedures Involving Pesticides." In U. S. Congress, House, Committee on Government Operations. *Deficiencies in FIFRA*. Pp. 141–180.

———. "Need to Resolve Questions of Safety Involving Certain Registered Uses of Lindane Pesticide Pellets." Included in U. S. Congress, House, Committee on Government Operations. *Deficiencies in FIFRA*. Pp. 181–217.

Graham, Frank, Jr. *Since Silent Spring*. Boston: Houghton Mifflin, 1970.

Gypsy Moth, Record Group 7, National Archives; full title is "Correspondence Relating to Investigations of Gypsy and Brown-Tail Moths, 1905–1908." Record Group 7, National Archives.

"The Gypsy and Brown-Tail Moths." Bulletin 1, Office of the Superintendent for Suppressing the Gypsy and Brown-Tail Moths. Boston: Wright and Potter, state printers, 1905.

Haller, H. L. and Ruth L. Busby. "The Chemistry of DDT." In U. S. Department of Agriculture. *Science in Farming, Yearbook of*

Agriculture for 1943–1947. Washington: Government Printing Office, 1947. Pp. 616–622.

Haller, H. L. and Stanley J. Cristol. "The Development of New Insecticides." In U. S. Office of Scientific Research and Development. *Advances in Military Medicine*. Vol. 2, Pp. 621–626.

Hamerstrom, Frank N., Jr. "An Ecological Appraisal of the Peregrine Decline." In Hickey, *Peregrine Falcon Populations*. Pp. 509–513.

Harding, T. Swann. "How much Poison can we eat?" *Scientific American* 149 (November 1933), 197–199.

Harris, C. R. S. *Allied Military Administration of Italy 1943–1945*. London: Her Majesty's Stationary Office, 1957.

Hart, Larry G., Robert W. Shultice, and James R. Fouts. "Stimulatory Effects of Chlordane on Hepatic Microsomal Drug Metabolism in the Rat." *Toxicology and Applied Pharmacology* 5 (May 1963), 371–386.

Hart, Larry G., and James R. Fouts. "Effects of Acute and Chronic DDT Administration on Hepatic Microsomal Drug Metabolism in the Rat." *Proceedings of the Society for Experimental Biology and Medicine* 114 (November 1963), 388–392.

Hartzell, Albert and Frank Wilcoxon. "Analysis of Sprayed Apples for Lead and Arsenic." *Journal of Economic Entomology* 21 (February 1928), 125–130.

Hayes, Wayland, J., Jr. "The Effect of Known Repeated Oral Doses of Chlorophenethane (DDT) in Man." *Journal of the American Medical Association* 162 (27 October 1956), 890–897.

Helms, Douglas. "Just Lookin' for a Home: The Cotton Boll Weevil and the South." Ph.D. dissertation, Florida State University, 1977.

Henkin, Harmon, Martin Merta, and James Staples. *The Environment, the Establishment, and the Law*. Boston: Houghton Mifflin, 1971.

Hickey, Joseph J., ed., *Peregrine Falcon Populations—Their Biology and Decline*. Madison: University of Wisconsin Press, 1969.

Hickey, Joseph J. and Daniel W. Anderson. "The Peregrine Falcon: Life History and Population Literature." In Hickey. *Peregrine Falcon Populations*. Pp. 3–44.

Hickey, Joseph H., and L. Barrie Hunt. "Initial Songbird Mortality Following a Dutch Elm Disease Control Program." *Journal of Wildlife Management* 24 (July 1960), 259–265.

Hickey, Joseph J., J. A. Keith, and Francis B. Coon. "An Exploration of Pesticides in a Lake Michigan Ecosystem." *Pesticides in
the Environment and Their Effect on Wildlife. Journal of Applied
Ecology* 3 (supplement, 1966, 141–153.

Hickey, Joseph J. and James E. Roelle. "Conference Summary and
Conclusions." In Hickey. *Peregrine Falcon Populations*, Pp. 565–
566.

Higdon, Hal. "Obituary for DDT (in Michigan)." *New York Times
Magazine* (6 July 1969), 6.

Hilton, James G. (chairman, DDT Advisory Committee). "Report
of the DDT Advisory Committee to William D. Ruckelshaus,
Administrator, Environmental Protection Agency." 9 September 1971.

Hinds, W. E. "Boll Weevil Control for Louisiana, 1924." Extension
circular 71, Louisiana State University and Agricultural and
Mechanical College—Division of Agricultural Extension.
Baton Rouge, 1924.

———. "Killing Over-Wintering Weevils: the Pre-Square Poison-
ing." Extension circular 72, Louisiana State University and
Agricultural and Mechanical College—Division of Agricultural
Extension. Baton Rouge, 1924.

———. "Machines and Poison for Weevil Control." Extension
circular 73, Louisiana State University and Agricultural and
Mechanical College—Division of Agricultural Extension.
Baton Rouge, 1924.

———. "Poisoning the Boll Weevil on Fruiting Cotton: The
Regular Dusting Treatment." Extension circular 74, Louisiana
State University and Agricultural and Mechanical College—
Division of Agricultural Extension. Baton Rouge, 1924.

Hotchkiss, Neil, and Richard H. Pough. "Effect on Forest Birds of
DDT used for Gypsy Moth Control in Pennsylvania." *Journal of
Wildlife Management* 10 (July 1946), 202–207.

Houston, David F. "Study and Investigation of Boll Weevil and Hog
Cholera Plagues." House Document 463, U.S . Congress,
House of Representatives, 63rd Congress, 2nd Session. Wash-
ington: Government Printing Office, 1913.

Howard, Leland O. "Danger of Importing Insect Pests." In U. S.
Department of Agriculture. *Yearbook of the United States Depart-
ment of Agriculture, 1897*. Washington: Government Printing
Office, 1898.

———. "The Education of the Entomologists in the Service of U. S. Department of Agriculture." *Journal of Economic Entomology* 7 (June 1914), 274–280.

———. *Fighting the Insects: The Story of an Entomologist.* New York: Macmillan, 1933.

———. *A History of Applied Entomology.* Smithsonian Miscellaneous Collections. Vol. 84. Washington: Smithsonian Institution, 1930.

———. *The Insect Menace.* New York: Century, 1931.

———. "The Organization Meeting of the Association of Economic Entomologists at Toronto, August 1889." *Journal of Economic Entomology* 15 (February 1922), 26–30.

———. "The Practical Use of the Insect Enemies of Injurious Insects." In U. S. Department of Agriculture. *Yearbook of Agriculture, 1916.* Washington: Government Printing Office, 1916. Pp. 273–278.

———. "Progress in Economic Entomology in the United States." In U. S. Department of Agriculture. *Yearbook of Agriculture, 1899.* Washington: Government Printing Office, 1899. Pp. 135–156.

Howard, Leland O., and W. F. Fiske. "The Importation into the United States of the Parasites of the Gypsy Moth and the Brown-Tail Moth." Bulletin 91, USDA Bureau of Entomology. Washington: Government Printing Office, 1911.

Hunt, Eldridge G. "Pesticide Residues in Fish and Wildlife of California." In Hickey. *Peregrine Falcon Populations,* Pp. 455–460.

Hunt, Eldridge G. and A. I. Bischoff. "Inimical Effects on Wildlife of Periodic DDD Applications to Clear Lake." *California Fish and Game* 46 (January 1960), 91–106.

Hunt, L. Barrie. "Songbird Breeding Populations in DDT-Sprayed Dutch Elm Disease Communities." *Journal of Wildlife Management* 24 (April 1960), 139–146.

Hunter, W. D. "The Control of the Boll Weevil." USDA Farmers' Bulletin 216. Washington: Government Printing Office, 1905.

———. "Methods of Controlling the Boll Weevil." USDA Farmers' Bulletin 163. Washington: Government Printing Office, 1903.

———. "The Use of Paris Green in Controlling the Cotton Boll Weevil." USDA Farmers' Bulletin 211. Washington: Government Printing Office, 1904.

Hunter, W. D. and B. R. Coad. "The Boll Weevil Problem." USDA Farmers' Bulletin 1329. Washington: Government Printing Office, 1933.

Hunter, W. D. and W. D. Pierce. *"The Mexican Cotton Boll Weevil: A Summary of the Investigation of this Insect up to December 31, 1911."* Bulletin 114, USDA Bureau of Entomology. Washington: Government Printing Office, 1912.

Insight Team of the Sunday Times of London. *Suffer the Children: The Story of Thalidomide.* New York: Viking, 1979.

Jackson, Charles O. *Food and Drug Legislation in the New Deal.* Princeton: Princeton University Press, 1970.

Kleinert, Stanton J., Paul E. Degurse, Thomas L. Wirth, and Linda C. Hall. *DDT and Dieldrin Residues Found in Wisconsin Fishes from the Survey of 1966.* Research report 23, Fisheries. Madison: Wisconsin Conservation Department, Research and Planning Division, 1967.

Knight, Paul. "The Development and Present Status of Entomological Courses in American Colleges and Universities." *Journal of Economic Entomology* 21 (December 1928), 871–877.

Knipling, Edward F. "The Greater Hazard—Insects or Insecticides." *Journal of Economic Entomology* 46 (February 1953), 1–7.

————. "DDT Insecticides Developed for Use by the Armed Forces." *Journal of Economic Entomology* 38 (February 1945), 201.

Kulp, J. Laurence, Walter R. Eckelmann, Arthur R. Schulert. "Strontium-90 in Man." *Science* 125 (8 February 1957), 219–225.

Lamb, Ruth deForest. *American Chamber of Horrors: The Truth about Food and Drugs.* New York: Farrar and Rinehart, 1936.

Laug, Edwin P., Arthur O. Nelson, O. Garth Fitzhugh, and Frieda M. Kunze. "Liver Cell Alteration and DDT Storage in the Fat of the Rat Induced by Dietary Levels of 1 to 50 ppm DDT." *Journal of Pharmacology and Experimental Therapeutics* 98 (March 1950), 268–273.

Laug, Edwin P., Frieda M. Kunze, and C. S. Prickett. "Occurrence of DDT in Human Fat and Milk." *A. M. A. Archives of Industrial Hygiene and Occupational Medicine* 3 (March 1951), 245–246.

Lauger, P., H. Martin, and P. Muller. "Über Konstitution und toxische Werkung von natürlichen und neuen synthetischen insektentötenden Stoffen." *Helvitica Chimica Acta* 27 (June 1945), 892.

Leary, James C., William J. Fishbein, and Lawrence C. Salter. *DDT and the Insect Problem*. New York: McGraw-Hill, 1946.

Leopold, Aldo. *A Sand County Almanac*. New York: Ballantine, 1970.

Libby, W. F. "Radioactive Fallout and Radioactive Strontium." *Science* 123 (20 April 1956), 657–662.

Linton, Fred B. "Leaders in Food and Drug Law." *Food Drug Cosmetic Law Journal* 5 (April 1950), 103–115; 5 (June 1950), 326–339; 5 (August 1950), 479–493; 5 (November 1950), 771–787.

Loucks, Orie L. "The Trial of DDT in Wisconsin." In J. Harte and R. H. Socolow, eds., *Patient Earth*. New York: Holt, Rinehart & Winston, 1971. Pp. 88–107.

Lyle, Clay. "Achievements and Possibilities in Pest Eradication." *Journal of Economic Entomology* 40 (February 1947), 1–8.

McDonald, R. E. "The Boll Weevil: A Review of the Methods of Control." Texas Department of Agriculture, Bulletin 74. Austin: Texas Department of Agriculture, 1923.

MacIntyre, Angus A. "The Politics of Nonincremental Domestic Change: Major Reform in Federal Pesticide and Predator Control Policy." Ph.D. dissertation, University of California, Davis, 1980.

McLean, Louis A. "Pesticides and the Environment." *Bioscience* 17 (September 1967), 613–617.

McNeil, William H. *Plagues and People*. New York: Doubleday, 1975.

Mallis, Arnold. *American Entomologists*. New Brunswick: Rutgers University Press, 1971.

Marlatt, C. L. "Boll Weevil Notes, April 24, to May 6, 1896." Typewritten copy in Townsend Correspondence, SFCII, Record Group 7, National Archives.

Martin, Calvin. *Keepers of the Game: Indian-Animal Relationships and the Fur Trade*. Berkeley: University of California Press, 1978.

Massachusetts State Board of Agriculture. *Agriculture in Massachusetts*. Annual report of the Secretary of the Massachusetts State Board of Agriculture. Boston: state printer, annual.

Medford [*Massachusetts*] *Mercury*. *Ravages of the Gypsy Moth in Medford*. Medford: *Medford Mercury*, 1905. Copy in Gypsy Moth, Record Group 7, National Archives.

Moore, Ernest G. *The Agricultural Research Service*. New York: Praeger, 1967.

Moore, William. "The Need of Chemistry for the Student of Entomology." *Journal of Economic Entomology* 16 (April 1923), 172–176.

Morison, Elting E., ed. *Letters of Theodore Roosevelt*. Cambridge: Harvard University Press, vol. 6, 1952; vol. 7, 1954.

Myers, C. N., Binford Throne, Florence Gustafson, and Jerome Kingsbury. "Significance and Danger of Spray Residue." *Industrial and Engineering Chemistry* 25 (1933), 624–628.

"NACA Speakers Emphasize Industry's Role in Pesticide Safety." *Agricultural Chemicals* 18 (October 1962), 20.

National Research Council. *Pest Control and Wildlife Relationships*, part 1; *Evaluation of Pesticide-Wildlife Problems*, part 2; *Policy and Procedures for Pest Control*, part 3. Publications 920-A and 920-B of the National Research Council. Washington: National Research Council, 1962.

Neal, Paul A., W. F. Van Oettingen, R. C. Dunn, and N. E. Sharpless. "Toxicity and Potential Dangers of Aerosols and Residues from Such Aerosols Containing Three Percent DDT." Supplement 183 to *U. S. Public Health Service Reports*. Washington: Government Printing Office, 1945.

Neal, Paul A., W. F. Van Oettengen, W. W. Smith, R. B. Malmo, R. C. Dunn, H. E. Moran, T. R. Sweeney, D. W. Armstrong, and W. C. White, "Toxicity and Potential Dangers of Aerosols, Mists, and Dusting Powders containing DDT." Supplement 177 to *U. S. Public Health Service Reports* Washington: Government Printing Office, 1944.

Neal, Paul A., Waldemar C. Dressen, Thomas I. Edwards, Warren H. Reinhart, Stewart H. Webster, Harold T. Castenberg, and Lawrence T. Fairhill. "A Study of the Effect of Lead Arsenate Exposure on Orchardists and Consumers of Sprayed Fruit." U. S. Public Health Service Bulletin 267. Washington: Government Printing Office, 1941.

"The Non-Ban on DDT—A Federal Phase-Out Fizzle." *Audubon Magazine* 74 (July 1971), 93.

O'Dell, W. V. "The Gypsy Moth Control Program." *Journal of Forestry* 57 (April 1960), 271–273.

O'Kane, W. C., C. H. Hadley, Jr., and W. K. Osgood. "Arsenical Residues after Spraying." Bulletin 183 of the New Hampshire Agricultural Experiment Station, June 1917.

Ormsbee, Richard. "The Toxicology and Mechanism of Action of DDT in Mammals." In U. S. Office of Scientific Research and

Development, Committee on Medical Research. *Advances in Military Medicine*. Vol. 2.

Orr, L. W., and L. O. Mott. "The Effects of DDT Administered Orally to Cows, Horses, and Sheep." *Journal of Economic Entomology* 38 (April 1945), 418–432.

Perkins, John H. "The Quest for Innovation in Agricultural Entomology, 1945–1978." In David Pimentel and John H. Perkins, eds. *Environmental, Socioeconomic and Political Aspects of Pest Management Systems*. AAAS selected symposium no. 43 Boulder, Colorado: Westview Press, 1979.

————. "Reshaping Technology in Wartime: The Effect of Military Goals on Entomological Research and Insect-Control Practices." *Technology and Culture* 19 (1978), 169–186.

"Pesticides as Possible Factors Affecting Raptor Populations— Summary of a Round Table Discussion and Additional Comments by the Conferees." In Hickey. *Peregrine Falcon Populations*. Pp. 455–485.

"Pesticides and the Industry and Target of One-Sided Attack," *Agricultural Chemicals* 17 (August 1962), 20.

Phelan, R. K. "Problems in Manufacturing Baby Food." *Food Drug Cosmetic Law Journal* 6 (April 1951), 295–302.

Pierce, W. Dwight, R. A. Cushman and C. E. Hood. "The Insect Enemies of the Cotton Boll Weevil." Bulletin 100, USDA Bureau of Entomology. Washington: Government Printing Office, 1912.

Popham, W. L. and David G. Hall, "Insect Eradication Programs." *Annual Review of Entomology* 3 (1958), 335–354.

"Population Biology and Significance of Trends." In Hickey. *Peregrine Falcon Populations*, 542–552.

President's Science Advisory Committee. *Use of Pesticides*. Washington: Government Printing Office, 1963.

Princi, Frank. "Toxicology and Hazard Record of the Newer Pesticides." *Agricultural Chemicals* 7 (January 1952), 44.

Puleston, Dennis. "Defending the Environment—A Case History." Brookhaven lecture series, no. 104, September 15, 1971. Springfield, Virginia: National Technical Information Service, 1971.

Ratcliffe, Derek A. "The Status of the Peregrine in Great Britain." *Bird Study* 10 (June 1963), 56–90.

"Report of the Accomplishments of the Work on Cotton Insect Investigations of the Bureau of Entomology and Plant Quarantine and Preceding Organizations of the United States Department of Agriculture from 1854 to 1938." Typewritten manuscript, personal copy of Dr. Douglas Helms, National Resources Branch, National Archives.

"Report of the Committee on Relation of Entomology to Conservation." *Journal of Economic Entomology* 40 (February 1947), 149–150.

"Report of the Entomologist." In USDA *Annual Reports of the Department of Agriculture*. Washington: Government Printing Office, annual.

Robbins, Chandler S., Paul F. Springer and Clark G. Webster. "Effects of Five-Year DDT Application on Breeding Bird Populations." *Journal of Wildlife Management* 15 (April 1951), 213–216.

Rosenberg, Charles E. "The Adams Act: Politics and the Cause of Scientific Research." *Agricultural History* 38 (January 1964), 3–12.

———. "Science, Technology and Economic Growth: The Case of the Agricultural Experiment Station Scientist, 1875–1914." *Agricultural History* 45 (January 1971), 1–20.

Rudd, Robert L. *Pesticides and the Living Landscape*. Madison: University of Wisconsin Press, 1964.

Ruggles, A. G. "Pioneering in Economic Entomology." *Journal of Economic Entomology* 17 (February 1924), 34–41.

Sax, Joseph L. *Defending the Environment*. New York: Knopf, 1971.

Scott papers—private collection of papers of Walter E. Scott, former assistant to the secretary, Wisconsin Department of Natural Resources.

Shelton, Ronald Lee. "The Environmental Era: A Chronological Guide to Policy and Concepts, 1962–1972." Ph.D. dissertation, Cornell University, 1973.

Smith, Harry S. "What May We Expect from Biological Control?" *Journal of Economic Entomology* 16 (December 1923), 506–511.

Snowden, Mason. "Cotton Growing In Louisiana: Demonstration Methods Under Boll Weevil Conditions." Extension circular 18, Louisiana State University and Agricultural and Mechanical College—Division of Agriculture Extension. (Baton Rouge, 1916).

Sobelman, Max, comp. *Selected Statements from State of Washington DDT Hearings Held in Seattle, October 14, 15, 16, 1969.* Published by the DDT Producers of the United States (Diamond

Shamrock Corp., Lebanon Chemical Corp., Montrose Chemical Corp. of California, Olin Corp.) December 1970.

Soper, Fred L., W. A. Davis, F. S. Markham, and L. A. Riehl. "Typhus Fever in Italy, 1943–1945, and its Control with Louse Powder." *American Journal of Hygiene* 45 (May 1947), 305–334.

Southern Field Crop Insect Investigation, Record Group 7, National Archives.

Sprunt, Alexander, IV. "Bald Eagles Aren't Producing Enough Young." *Audubon Magazine* 65 (January–February 1963), 32–35.

————. "An Eagle-eyed Look at Our Bald Eagle." *Audubon Magazine* 63 (November–December 1961), 324–326.

Sutherland, W. W. "No More Legislation Needed." *Agricultural Chemicals* 7 (April 1952), 41–43.

Swingle, Roger U., Russell R. Whitten, and E. G. Brewer. "Dutch Elm Disease." In U. S. Department of Agriculture. *Trees: The Yearbook of Agriculture, 1949.* Washington: Government Printing Office, 1949. Pp. 451–452.

"Symposium—Toxicity of DDT." *Journal of Economic Entomology* 39 (April, 1946), 425.

Townsend, C. H. Tyler. "Report on the Mexican Cotton Square-and-Boll Weevil (*Anthonomus Grandis* Boheman) in Texas." 20 December 1894. Copy in Townsend Correspondence, Record Group 7, National Archives. Printed as "Report on the Mexican Cotton-Boll Weevil in Texas." *Insect Life* 7 (March 1895), 295–309.

Townsend Correspondence, Record Group 7, "Letters and Reports Received from C. H. T. Townsend, 1894–1898, Records of the Division of Southern Field Crop Insect Investigations." National Archives.

Ulyett, G. C. "Insects, Man and the Environment." *Journal of Economic Entomology* 44 (August 1951), 459–464.

U. S. Congress, House, Committee on Agriculture. *Federal Pesticide Control Act of 1971.* Hearings before the Committee on Agriculture, House, 92nd. Congress, 1st session, on H. R. 26, H. R. 1077, H. R. 1722, H. R. 4152, H. R. 4596, H. R. 5182, H. R. 6576, and H. R. 6761. Washington: Government Printing Office, 1971.

U. S. Congress, House, Committee on Government Operations. *Deficiencies in Administration of Federal Insecticide, Fungicide, and Rodenticide Act* (FIFRA). Hearings before a subcommittee of the Committee on Government Operations, 91st Congress, 1st session. Washington: Government Printing Office, 1969.

————. *Deficiencies in Administration of Federal Insecticide, Fungicide, and Rodenticide Act* (FIFRA). House Report 91-637, 91st Congress, 1st Session. Washington: Government Printing Office, 1969.

U. S. Congress, House, Select Committee to Investigate the Use of Chemicals in Food Products. *Chemicals in Food Products*. Hearings before the House Select Committee to Investigate the Use of Chemicals in Food Products, 81st Congress, 2nd Session. Washington: Government Printing Office, 1951.

————. *Chemicals in Food Products*. Hearings before the House Select Committee to Investigate the Use of Chemicals in Food Products, 82nd Congress, 1st Session. Washington: Government Printing Office, 1953.

U. S. Congress, House, Subcommittee on Forests of the Committee on Agriculture. *Permit the Use of DDT*, Hearings before the Subcommittee on Forests of the Committee on Agriculture on H. R. 10796, 93rd Congress, 1st Session. Washington: Government Printing Office, 1974.

U. S. Congress, Senate, Committee on Agriculture and Forestry. *Pest Control Research*. Hearings before the Subcommittee on Agricultural Research and General Legislation, Committee on Agriculture and Forestry, Senate, on S. 1794. Washington: Government Printing Office, 1971.

U. S. Congress, Senate, Committee on Commerce. *Foods, Drugs, and Cosmetics*. Hearings before the Committee on Commerce, U. S. Senate, on S. 2800, 73rd Congress, 2nd Session. Washington: Government Printing Office, 1934.

U. S. Congress, Senate, Committee on Government Operations. *Pesticides and Public Policy*. Report of the Committee on Government Operations, U. S. Senate, 89th Congress, 2nd Session, report 1379. Washington: Government Printing Office, 1966.

U. S. Congress, Senate, Subcommittee of the Committee on Commerce. *Foods, Drugs, and Cosmetics*. Hearings before a subcommittee on Commerce, U. S. Senate, on S. 1944, 73rd Congress, 2nd Session. Washington: Government Printing Office, 1934.

————. *Foods, Drugs, and Cosmetics*. Hearings before a subcommittee of the Committee on Commerce, U. S. Senate, on S. 5, 74th Congress, 1st Session. Washington: Government Printing Office, 1935.

U. S. Congress, Senate, Subcommittee on Agricultural Research and General Legislation of the Committee on Agriculture and Forestry. *Federal Environmental Pesticide Control Act*. Hearings before the Subcommittee on Agricultural Research and General Legislation of the Committee on Agriculture and Forestry, 92nd Congress, 2nd Session. Washington: Government Printing Office, 1971.

U. S. Congress, Senate, Subcommittee on Reorganization and International Organization of the Committee on Government Operations, *Interagency Coordination on Environmental Hazards (Pesticides)*. Hearings before the Subcommittee on Reorganization and International Organization of the Committee on Government Operations, U. S. Senate, 88th Congress, 1st Session. Washington: Government Printing Office, 1964. Report titled *Pesticides and Public Policy*, report 1379, 89th Congress, 2nd Session.

U. S. Department of Agriculture. *Insects, The Yearbook of Agriculture for 1952*. Washington: Government Printing Office, 1952.

———. *Annual Report of the United States Department of Agriculture*, 1868. Washington: Government Printing Office, 1869.

U. S. Department of Commerce, Bureau of the Census. *Abstract of Census of Manufactures, 1914*. Washington: Government Printing Office, 1917.

———. *Biennial Census of Manufactures, 1937*. Part 1. Washington: Government Printing Office, 1939.

———. *Census of Manufactures: 1947—Produce Supplement*. Washington: Government Printing Office, 1950.

———. *U. S. Census of Manufactures: 1958, Volume II, Industry Statistics, Part 1, Major Groups 20–28*. Washington: Government Printing Office, 1961.

———. *Census of Manufactures: 1967, Volume II, Industry Statistics, Part 2, Major Groups 25–33*. Washington: Government Printing Office, 1971.

U. S. Department of Health, Education, and Welfare. *Report of the Secretary's Commission on Pesticides and Their Relationship to Environmental Health*. Washington: Government Printing Office, 1969.

U. S. Department of the Interior, Geological Survey. *The National Atlas of the United States*. Washington: Government Printing Office, 1970.

U. S. Office of Scientific Research and Development, Committee on Medical Research. *Advances in Military Medicine*. Vol. 2. Boston: Little, Brown, 1948.

University of Wisconsin Department of Entomology. "Critique of 'Insecticides and People.' " Unpublished and privately circulated critique of a panel discussion presented on radio station WHA, 7 November 1962.

Valiga v. National Food Co. 58 Wis 2d 232, decided 20 April 1973.

van den Bosch, Robert. *The Pesticide Conspiracy.* New York: Doubleday, 1978.

Wallace, George J. "Another Year of Robin Losses on University Campus." *Audubon Magazine* 62 (March–April 1960), 67.

———. "Insecticides and Birds." *Audubon Magazine* 61 (January–February 1959), 10.

Wallace, George J., Walter P. Nickell, and Richard F. Bernard. *Bird Mortality in the Dutch Elm Disease Program in Michigan.* Bulletin 41 of Cranbrook Institute of Science, Bloomfield Hills, Michigan. Bloomfield Hills, Michigan: Cranbrook Institute of Science, 1961.

Waller, Wilhelmine Kirby. "Poison on the Land." *Audubon Magazine* 60 (March–April 1958), 68.

Wheeler, Charles M. "Control of Typhus in Italy 1943–1944 by the Use of DDT." *American Journal of Public Health* 36 (February 1946), 119–129.

Whitaker, Adelynne Hiller. "A History of Federal Pesticide Regulation in the United States to 1947." Ph.D. dissertation, Emory University, 1974.

White-Stevens, Robert H. "Communications Create Understanding." *Agricultural Chemicals* 17 (October 1962), 34.

Whitten, Jamie L. *That We May Live.* Princeton: D. Van Nostrand, 1966.

Whitten, Russell R. and Roger U. Swingle. "The Dutch Elm Disease and Its Control." USDA Agriculture Information Bulletin 193. Washington: Government Printing Office, 1958.

Whorton, James C. "Insecticide Residues on Foods as a Public Health Problem: 1865–1938." Ph.D. dissertation, University of Wisconsin, 1969.

———. *Before Silent Spring.* Princeton: Princeton University Press, 1974.

Wiley, Harvey W. *The History of a Crime against the Food Law.* Washington: Harvey W. Wiley, 1929.

———. "The Poison on Our Fruits." *Good Housekeeping* 87 (July 1928), 100.

Wisconsin Department of Agriculture and the University of Wisconsin. "Dutch Elm Disease Control Recommendations for 1969." One page, mimeographed copy from Scott papers.

Wisconsin Department of Natural Resources. "Frederick M. Baumgartner et al. v. City of Milwaukee and Buckley Tree Service." Docket 4C-68-5-1, hearing held 18 October 1968. Typescript, Wisconsin Department of Natural Resources, 1968. Copy from Scott papers.

———. "In the Matter of the Petition of CNRA of Wisconsin, Inc. et al. for a Declaratory Ruling on DDT." Docket 3-DR-1. Typescript, Wisconsin Department of Natural Resources, 1969. Copy from Van Susteren, hearing examiner.

Stafford, Willard S. "Exceptions to Examiner's Statement of Statutes, Rules, and Issues." No date.

Van Susteren, Maurice H. "Examiner's Ruling on Motion of Industry Task Force for DDT of the National Chemical Association in Support of Motion to Dismiss Proceedings." 16 November 1969. Copy from Scott papers.

———. "Examiner's Summary of Evidence and Proposed Ruling." Copy from Scott papers.

Waiss, Frederick S. and Willard S. Stafford. "Exceptions of Intervenor Industry Task Force for DDT of the National Agricultural Chemicals Association to the Examiner's Summary of Evidence and Proposed Ruling." Brief submitted in opposition to the examiner's proposed ruling in the DDT petition by CNRA. Copy from Scott papers.

Wisconsin Department of Natural Resources. "Order Affirming and Adopting Examiner's Proposed Ruling." 23 October 1970. Copy from Scott papers.

Wisconsin Legislative Council Staff. "Regulation of DDT and other Pesticides: Recent Developments in Wisconsin." Research bulletin 5, September, 1969. Copy from Scott papers.

Wisconsin State Broadcasting Service. "Summary of Broadcasts Dealing with *Silent Spring* by Rachel Carson and the Subject of Pesticides." Typescript, Wisconsin State Broadcasting Service, 1962. Copy from Aaron Ihde.

Woodwell, George M., Charles F. Wurster, Jr., and Peter A. Isaacson. "DDT Residues in an East Coast Estuary: A Case of Biological Concentration of a Persistent Insecticide." *Science* 156 (12 May 1967), 821–824.

Woodworth, C. W. "The Insecticide Industry in California." *Journal of Economic Entomology* 5 (August 1912), 358–364.

Worrell, Albert C. "Pests, Pesticides, and People." *American Forests* 66 (July 1960), 39–81.

Worster, Donald. *Nature's Economy*. Sierra Club Books, 1977; New York: Anchor, 1979.

Wurster, Doris H., Charles F. Wurster, Jr., and Walter N. Strickland. "Bird Mortality Following DDT Spray for Dutch Elm Disease." *Ecology* 46 (early Summer 1965), 488–499.

Yannacone et al. v. Dennison et al., 285 N. Y. S. 2a 476. New York.

Yuill, J. S., D. A. Islaer, and George D. Childress. "Research on Aerial Spraying." In U. S. Department of Agriculture. *Insects*. Pp. 252–258.

Zeidler, Othmar. "Verbindung von Chloral mit Brom- und Chlorbenzol." In "Untersuchen uber die Synthetische Darstellung von Aromatischen Verbindungen durch Wasserentziehung." *Berichte der deutschen Chemischen Gesellschaft zu Berlin* 8 (1874), 1180–1181.

Zimmerman, O. T. and Irvin Levine. *DDT: Killer of Killers*. Dover, N.H.: Industrial Research Service, 1946.

Zinsser, Hans. *Rats, Lice, and History*. Boston: Little, Brown, 1935.

INDEX

Howard, Leland O., 19, 33, 37
Howell, Joseph C., 96
H. P. Cannon Company, 212
Hunt, Eldridge G., 121, 133
Hunt, Reid, 45
Hunt Commission, 45, 48

Illinois Audubon Society, 85
Iltis, Hugh H., 109, 162
Industry Task Force for DDT (of the
    NACA), 158
insect-borne disease, 60–63
insect control: methods, 19–20, 29;
    obstacles, 24–25
insect ecology, 33
Insect Life, 22
Insect Menace, The, 37
Insecticide Act of 1910, 24, 35, 251
insecticide manufacturers, 36, 71–74,
    106, 143
insecticide residues: analysis of, 135;
    and avian reproduction, 168,
    175–176; bioconcentration of, 137;
    effects on food, 40–55, 66; effects
    over time, 52; effects on wildlife,
    121; environmental accumulation,
    117; environmental monitoring,
    90, 124, 125, 138; environmental
    study, 129–140; and human
    health, 39–55, 64, 88, 180; in
    mammals, 49, 138; criteria for
    safety, 40, 53, 70, 120–121, 174,
    180–182, 193; need for study, 139;
    and thin eggshells, 139, 175–176;
    tolerances, 6, 43, 46, 53, 255; units
    of measurement, 246. *See also*
    DDT
insecticides: advantages of, 142; de-
    velopment of, 17–38; effect on
    human health, 39–55, 64, 88, 180;
    federal regulation of, 43–55, 75,
    114, 200–203, 207–210, 235–238;
    manufacture of since World War
    II, 73; production of, 251
"Insecticides and People," 109

insects, spread of, 18
Integrated Pest Management, 13,
    224, 242
International Apple Association, 48,
    51, 69

Jardine, William, 44
Johnson, Howard D., 151
Journal of Economic Entomology, 23

Kedzie, R. C., 40
Kehoe, Robert, 67
Keith, James, 121, 221
Kelsey, Frances, 104
Kennedy, John F., 101
Kettering Laboratory, 61, 67
Kirk, Alan, 220
Kirk, Norman J., 62
Kirkland, A. J., 40
Knipling, Edward, 69, 72, 243
Koebele, Albert, 31

Lake Michigan, 137–138
Lamb, Ruth, 49
Larrick, George, 121
Laug, Edwin, 65
Laws, Edward, 214
lead: on food, 39–55; poisoning, 50.
    See also insecticide residues
Leopold, Aldo, 155
Lofroth, Goran, 186–187
Loucks, Orie, 155, 162, 195
Lovenhart, Arthur S., 45–46
Lyle, Clay, 77
Lynn, George, 118

Macek, Kenneth, 173, 219
McClosky, Pete, 199
McConnell, Robert, 160, 186
McLean, Louis A., 158, 181; at Wis-
    consin hearing, 160–175
MacMullan, Ralph, 149
Malignant Neglect, 239
Maren, Thomas, 217
Marvin, Phillip, 191

Rotrosen, Samuel, 144
Ruckelshaus, William D., 231
Rudd, Robert, 7, 101, 157, 176

safety of residues, *see* insecticide res-
idues
Saffiotti, Umberto, 215
Sax, Joseph, 198
Schmitt, Karl, 109
Science Student Union, 172
scientists in public affairs, 5, 97, 123,
198–199, 242–245
Sierra Club, 206
Silent Spring, 3–4, 7, 75–76, 78,
98–99, 121, 129, 142, 152, 160,
176, 197–198, 201, 235; attacks on,
105–113; effect on public, 76,
97–99, 197
Sindelar, Charles R., Jr., 131–132
Smolker, Robert, 144
spruce budworm, 76
Sprunt, Alexander IV, 96, 134
Stafford, Willard S., 172, 179, 186–
191
Stahr, Elvis J., 205
Steinbach, Allan B., 192
Stickel, Lucille F., 133, 174
Stickel, William H., 133, 219
strontium-90, 102–103, 186
"Sue the Bastards," 147
Suffolk County Mosquito Control
Commission, 144
Sutherland, W. W., 71
Sweeney, Edward, 212, 220–223,
228; decision on DDT, 231–232
Switzer, Bruce, 226
systems analysis, 195

Taromina, Anthony, 144
technology, public faith in, 102, 106,
112, 143, 160, 235
thalidomide, 104
That We May Live, 142
thin eggshells, 132–133, 137–139; in
pelicans, 221; testimony at EPA
hearing, 217, 221, 222, 226–227;

testimony at Wisconsin hearing,
169, 175–176
Thomas, Cyrus, 22
Tomatis, Lorenzo, 215
total population management, 242
Townsend, C. H. Tyler, 26–29
Tracy, Robert E., 177
Trovelot, Leopold, 32
Tugwell, Rexford G., 47
Typhus, in Naples, 3, 61–62

Udall, Stewart, 121, 209
Ullyet, G. C., 167
United States Department of Agriculture, 13, 191; conflict of interest, 42; and DDT defense, 213,
217; and DDT registration, 207,
211; defends pesticide regulation,
69, 182, 206; at EPA hearing,
210–230; regulation of pesticides,
8, 114, 117–118, 163, 200–202, 206
United States Entomological Commission, 20
United States Typhus Commission,
62
"Use of Pesticides," 113

van den Bosch, Robert, 157, 166,
244
Van Susteren, Maurice, 152, 160,
232
vapor phase chromatography
(VPC), 14, 135, 168, 187, 257
Velsicol Chemical Corporation, 158

Waiss, Frederick S., 172, 188–189
Wallace, George, 83, 97, 151, 191
Wallace, Henry, 48
War Food Administration, 60
war on insects, 36
War Production Board, 63
Waterman, Charles, 44
Weisner, Jerome, 133
Welch, Richard, 172–175
West Michigan Environmental Ac-
tion Council, 206